Maps and the Writing of Space in Early Modern England and Ireland

Maps and the Writing of Space in Early Modern England and Ireland

Bernhard Klein
Lecturer in English
University of Dortmund
Germany

Published by
PALGRAVE MACMILLAN
Houndmills, Basingstoke, Hampshire RG21 6XS and
175 Fifth Avenue, New York, N.Y. 10010
Companies and representatives throughout the world

PALGRAVE MACMILLAN is the global academic imprint of the Palgrave Macmillan division of St. Martin's Press, LLC and of Palgrave Macmillan Ltd. Macmillan® is a registered trademark in the United States, United Kingdom and other countries. Palgrave is a registered trademark in the European Union and other countries.

ISBN-13: 978–0–333–77933–0
ISBN-10: 0–333–77933–9

This book is printed on paper suitable for recycling and made from fully managed and sustained forest sources. Logging, pulping and manufacturing processes are expected to conform to the environmental regulations of the country of origin.

A catalogue record for this book is available from the British Library.

Library of Congress Catalog Card Number: 00–033348

Printed and bound in Great Britain by
CPI Antony Rowe, Chippenham and Eastbourne

To the memory of my father
Heinz Werner Klein
1927–1980

Contents

List of Figures

List of Plates

Acknowledgements

As this study crawled at snail's pace from enthusiastic early research through mid-thesis crisis towards its final form, I incurred many debts to teachers, colleagues and friends without whose generous help and expert advice I could never have completed this project. The book started life as a PhD thesis at Frankfurt University where in 1998 it was passed by Klaus Reichert, Eckhard Lobsien (both Frankfurt), and Manfred Pfister (Berlin). In the early stages, many people read hopelessly inadequate draft chapters and kept a straight face; thanks are due to Catarina Albano, Tom Healy, Gesa Mackenthun, Philip Orr, Quentin Skinner and the members of the Frankfurt Early Modern Research Group in the academic year 1994/5: Ina Habermann, Christian Schmitt, Susanne Scholz and Peter Sillem. To Gesa I am indebted in more ways than she may realize; without the example of her unfailing enthusiasm and down-to-earth approach to academic life I may never have had the courage to get started on this project. I am particularly grateful to those who went through the ordeal of reading the typescript in full: Andrew Gordon, Ina Habermann, David Matthews and Andrew McRae. All offered valuable criticism, helped me improve my English, and saved me from numerous blunders. David made sure that food for thought did not remain a hungry metaphor, both in London and France. Special thanks must go to Andrew Gordon who listened patiently when ideas were raw and who proved an ingenious partner in running a conference; I would like to thank him and all the speakers and attendants at our 1997 *Paper Landscapes* conference in London for making that event such a rare academic treat. Many other people have given generously of their time and experience in discussions and informal conversations; I would like to thank especially the late Francis Barker, Victoria von Flemming, Anne Fogarty, John Gillies, Richard Helgerson, Alf Mentzer, Andrew Murphy, Manfred Pfister, Kurt Tetzeli and the members of the London Renaissance Seminar. I am grateful to Charmian Hearne for her initial interest in the project and to Eleanor Birne and Anne Rafique for seeing the typescript through the press. Earlier versions of several chapters were first read as conference papers to audiences in Vienna, Reading, Southampton, Dresden, Weimar, Dublin, Mainz and Frankfurt; I am grateful to everybody who offered helpful criticism on these occasions. Needless to say, since my stubbornness kept me from accepting all the good advice I was offered, any errors that remain are entirely my own responsibility.

Portions of several chapters have already appeared in print. I am grateful to the editors of the *Journal for the Study of British Cultures*, the *Irish University Review* and *Early Modern Literary Studies* for allowing me to re-use material, all of which appears here in revised form. A version of Chapter 8 is to appear in *Literature, Mapping, and the Politics of Space in Early Modern Britain*, edited by Andrew Gordon and myself. Permissions to reproduce illustrations have been granted by the British Library (Figures 2.1, 2.2, 2.3, 2.4 and 8.1; Plates 4, 5, 6, 7, 8, 9 and 11), the Stadt- und Universitätsbibliothek Frankfurt am Main (Figures 1.1, 1.2, 1.4, 1.5, 1.6 and 1.7; Plates 2 and 3), the Deputy Keeper of Records, Public Record Office of Northern Ireland (Plates 14, 15 and 16), the Council of Trustees, National Library of Ireland (Plates 12 and 13), the Städelsches Kunstinstitut, Frankfurt (Plate 1), the Public Record Office, Kew (Plate 10), and the Senckenbergische Bibliothek, Frankfurt (Figure 1.3). Shelf-marks are quoted in the list of illustrations.

This is a book on space which has benefited immensely from the academic, social and institutional spaces in which it was researched and written, reading rooms in English, Irish and German libraries, flats in London, Frankfurt, Munich and Dortmund, and university departments in England and Germany. One social space has been more stimulating than all the others, the old North Library in the British Museum, an exceptionally fruitful academic environment that has now sadly vanished forever. Laptop Alley is sorely missed by all its former residents. For the academic year 1995/6 I was granted a scholarship by the German Academic Exchange Service (DAAD) which enabled me to spend a year at Birkbeck College, University of London. This visit proved crucial for the development of my work, and I would like to thank the DAAD, Birkbeck College, and especially Tom Healy for having made it possible, as well as my former colleagues at the Frankfurt English department for granting me a year's leave of absence. My current colleagues in Dortmund have all been exceptional in their support and encouragement, especially Jürgen Kramer. The greatest debt of all, both academic and personal, is owed to Ina Habermann, my most critical, encouraging and inspiring reader.

Introduction:
The Cartographic Transaction

Johann Vermeer's 1668 painting *The Geographer* (Plate 1), now in the Städel art gallery in Frankfurt, shows a solitary man inside a modestly furnished room. Half bent over a table, supporting himself with his left hand on a book while balancing a compass in his right, he stares reflectively into the middle distance. The geographer may be in a contemplative, even melancholic mood, fully absorbed in his worldly labours. Maybe he is simply looking out of the window, having just been interrupted in his work.[1] His searching gaze, mirrored by our own inquisitive eyes as we scan the painted surface, attracts our attention not least because of the strong link the painting establishes between the window on the left of the canvas, the only source of light, and the geographer's illumined face, set off against a sombre, shadowy background.[2] What is there to see on the other side of the window? Trees, houses, a scene in the street, people in the yard? We have no means of knowing. All *we* can see is the actual window, not the image it frames. For the figure in the painting, the window is the point of visual contact with the outside world but for us, the viewers, it hides what it reveals to him. What keeps us at a distance from the world beyond the privacy of the room is the immaculate handling of linear perspective, skilfully employed to force the window into such an acute angle that only the carefully arranged play of shadows on the back wall reminds us of the exterior world encompassing the domestic scene.[3] Yet whatever the man in the centre of the room may be looking at, or brooding about, with such apparent intensity, it is part of the visual power of the painting, part of the representational trick played on the viewer, that we ask ourselves precisely this question: what has caught the man's attention?

Here is one *possible* answer to this question, one that takes its cue from the professional identity ascribed to the central figure in the painting.

As a geographer, an authority on the representation of space, the figure typifies a certain *kind* of interest in the physical world beyond the window, an interest clearly defined by the objects scattered across the canvas. The globe on top of the wardrobe, two rolled up maps on the floor, the sea chart of the Mediterranean on the back wall, the compass in his right hand, and what may well be another map on the table he is in the act of drawing,[4] all specify the technical and material range of this professional interest. In a sense, these objects re-import into the representational space of the canvas what the focus on the domestic takes away: the public sphere beyond the private study, mirrored on the inside by a range of geographical images taking precisely this outside world as their theme. The painting, in other words, is a meditation on the discipline of geography; and if the geographer's gaze is accepted as a further illustration of this main theme, it is one invested with a specific direction, focus and quality: directed at the surface of the physical landscape, it aims to translate the human experience of 'being in space' (the scene outside the window) into a set of material artefacts (map, globe) that offer to visually describe this space within the parameters of the ancient knowledge of geometry (signified by the compass). It is thus not without a certain inner logic that the geographer's immediate object of attention – the land itself – is hidden from our view and available only in its mediated form as a globe or map. The painting alerts us not to *what* the geographer sees, but to *how* he sees, to the historical and technical conditions shaping his perception of the world: an inquisitive 'view' inextricably linked to its articulation in the descriptive language of maps, charts and globes.

The dialectic of outside and inside may be a staple of the kind of Dutch genre painting to which *The Geographer* belongs but here serves quite a specific purpose: it divides experience from representation, referent from sign, in a move characteristic of modern cartography. On one level, this division is eminently practical. In the words of a contemporary English practitioner, there is no

> comparison in a poynt, lyne, angle, or measure wrought *in the fieldes and foule wether*, vnto the operations, framed *in a well lighted house*, vpon a faire leuell, and smooth table: the eye and hand hanging plumme ouer the worke, to be set down vpon Angles, measures, & all other regards curiously taken, and noted abroad.[5]

But on another level, the transfer of the map production process from an outside world of 'fieldes and foule wether' into the interior of the

'well lighted house' is not exclusively a matter of technical conveni-
ence. It almost casually reproduces a central strategy of the cartographic
image itself, a mode of visual representation perhaps best understood
as a highly artificial technology of signs invested with the unique power
to imitate in a network of lines and colours what we habitually refer to
as the 'real': it manages, without any apparent effort, to replace a natural
world beyond our physical control with the promise of mental order
wrapped in the Euclidean rhetoric of 'poynt, lyne, angle, or measure'.
My claim is thus not, or is at least only partially, that the portrait translates into an aesthetic composition the *empirical* process of map-making.
Rather, it represents what I take to be the *conceptual* stages of the cartographic transfer of world into map, articulated in the terms of a visual
language prevalent in seventeenth-century Holland.

These conceptual stages, I suggest, are threefold. First, space is *measured*. Preceding the scene depicted in the painting we may imagine a
land surveyor marching across a field, climbing the highest mountain
of the region, or demanding admittance to the local church steeple best
situated for taking accurate measurements of the surrounding landscape. With the practical help of compass, plane table or theodolite, and
the assistance of the specialized code of geometry, the space of the world
could be converted into a series of abstract mathematical figures. We
may thus spot in the gaze of Vermeer's geographer the etymological
origins of the word 'surveyor', literally an 'overseer'. Second, space is
visualized. With the measurements taken in the field, cartographers
retreated into the inner space of their study or workshop where – with
'the eye and hand hanging plumme ouer the worke' – they created
visual images that were to circulate in society as pictorial signifiers of
specific social, political or economic spaces: maps of estates, of the
nation, of the colonial world, etc. The map hanging on the back wall,
a sea chart of the Mediterranean, visualizes a space of multiple significance to the painting. As a navigational aid it makes accessible to Dutch
merchants a space of trade and commerce that ensures the material
availability of the objects making up the space represented as well as –
possibly – the space of representation itself: the cloth in the foreground,
or the textiles for the gown worn by the geographer, even the very
colours in which the canvas is painted, may have all been imported into
Holland via the maritime trade routes depicted in such charts. Third,
space is *narrated*. To grasp this dimension of Vermeer's painting we need
to reinsert it into the historical context from which it emerges, into the
first decades of a fragile Dutch independence gained from the Habsburg
monarch in 1648. It was the historical scenario of a newly sovereign

state that gave rise to the aestheticized scenes of domestic interiors celebrated in genre painting – scenes that in their emphasis on order and cleanliness were both images of the state and of the new, 'private' nature of the bourgeois Dutch Republic, a 'place apart from the "public" world of kings and courts'.[6] The domestic space 'narrated' by the painting is thus the pictorial correlate of a new political and economic order whose implicit reference is to the larger public space of the nation, surrounding both the painting itself and its productive context.

This line of enquiry may be pursued even further. For what the content of the picture plane suggests – the triple mechanism of spatial conversion through number, image and text – the painting repeats on the level of formal representation: the practical work of composing the image equally required a prior act of *measuring* to achieve the perfect balance of linear perspective;[7] the canvas as such is a further instance of *visualization* that draws more attention to itself than to the maps it includes on its surface; and its circulation in society as a material artefact generates a *narrative* of its own production that inevitably hinges on the skill and expertise of its painter – by general agreement, the modern 'master of light'. This self-reflexive structure points to analogies as well as differences between the material art object and its subject matter, the discourse of geography. For instance, to locate the common source of scale-map and linear perspective, as I have just implicitly done, in a practice of applied geometry (for which the suggestive grid of the window may well serve as a visual reminder) is not to argue universally for the existence of 'fixed viewpoints of perspective maps and paintings' that have 'shaped ways of seeing for four centuries'.[8] Such a proposition would seem to ignore a crucial distinction between maps and paintings, the different visual strategies they employ to address the observer: the kind of perspectival art practised by Vermeer directly incorporates the viewer in the space of the painting; by contrast, the physical landscape projected onto a map shuns linear perspective in favour of the geometric abstraction enabled by the cartographic grid. A scale-map aims not to hide the world from our view, like Vermeer's window, but to lay it open. Yet the common investment of map and painting in systems of applied geometry does testify to a shared background, if not to a 'fixed viewpoint' (not the most appropriate term for the decentred image of the scale-map) then more generally to a 'mathematization of experience', affecting the painting both as aesthetic sign and as material object. *The Geographer*, in other words, rehearses in subject and substance the complex discourses that linked art and science, painting and cartography, culture and commerce, throughout the early modern period.

The point of this reading is to offer Vermeer's portrait as an emblem for the wider pattern of cultural change that forms the topic of this study. More precisely, I would like to suggest that the pictorial fusion of the three discrete spatial operations that I have just extracted from *The Geographer* – measurement, visualization and narration – may be read, for the purposes of the analysis that follows, as the aestheticization of a shift in spatial awareness I will refer to as the 'cartographic transaction': the mental and material renegotiation of the lived space of experience *outside the window*, beyond the confines of the geographer's study – or, by metonymic extension, beyond the confines of the depopulated map. This hugely influential semantic transfer of organic into functional space – a process both produced by, and productive of, modern cartography – is enabled by a novel representational technology involving the mutually overlapping conceptual triad of number, image and text, which redefines in a movement of recognition and translation the social and political space of early modern Europe.

Such cartographic transactions are the theme of this study. My focus is more narrow than the continental scenario to which I have just alluded, limited spatially to England and Ireland, and temporally to the Elizabethan and Jacobean periods. Central to my enquiry are not the domestic spaces captured in many of Vermeer's paintings but, to stay within the image, the maps he often places on their walls. Throughout the chapters that follow, I will concentrate largely on the representation of an overt, frontal and intrinsically public world: the space of the globe, of the emerging nation, of the semi-colonial sphere of Ireland and of the English estate; in each case I am particularly interested in the representational strategies that serve to foreground the social and political dimensions of these spatial constructs. To impose structural coherence on my own text I have appropriated the conceptual triad outlined above and ordered my material as a sequence of three parts dealing respectively with 'Measurements', 'Cartographies' and 'Narratives'. These headings are not watertight categories; they are meant both to indicate the principal mechanisms of spatial transformation and to suggest the structural affinity between them. This does not imply that the acts of measuring, visualizing, and narrating space need to be thought of exclusively as a temporal sequence; they can clearly occur simultaneously, or be represented – as, for instance, in Vermeer's painting – within the discursive confines of a single aesthetic artefact. Still, it is the *suggestion* of an empirical process absorbed into a conceptual logic that makes this triadic structure a particularly suitable framing device for the critical readings that follow.

Vermeer's painting takes for granted that the viewer is no novice to the reading techniques appropriate to the representational sophistication of modern scale-maps. What this historical achievement requires as its enabling condition – the prior spread of map literacy – points to deeper cultural changes in European spatial consciousness. If the medieval conception of space relied on the notion of a 'hierarchic ensemble of places',[9] a cosmological order of things that structured space according to detailed systems of local meaning, the intellectual and practical forays of the Renaissance gradually helped bring into existence the concept of an infinite, open space with seemingly unlimited possibilities of expansion. Hence, the traditional symbolism of medieval cartography, where a map like the thirteenth-century Hereford *mappa mundi* functioned less as an index of place than as a visual record of divine creation, was confronted with a sense of geographic 'mimesis'. Such shifts in the mental organization of space are reflected by conceptual shifts in the notion of 'scale'. Frank Lestringant has recently suggested that in order to grasp the innovatory nature of the early modern geographical literature it does not suffice to resort either to the notion of 'exoticism' – which claims that the 'marvellous realities' of the classical world, recorded by Pliny, Solinus or Pomponius Melona, were 'gradually idealized or allegorized into new myths'[10] brought back from the Americas – or to the theory of the 'new horizons' – which argues that a narrow Mediterranean world opened up and slowly pushed forward its geographic boundaries, both east and west. Rather, the early modern spatial experience is best described as a 'sudden rupture of scales that changed people's way of viewing the world, and consequently the world itself',[11] a shift from a dominant spatial model in which the Mediterranean centre of the world was poised against a liminal, uncivilized and possibly non-human periphery, to a model which translated the earthly globe into a vast pictorial frame circumscribing a space ready for European inscription.

The chapters in Part I focus on one particular aspect of this 'rupture of scales', the social and political repercussions of such changes in spatial awareness. Each examines the confrontation between geographical space encoded as an inherently social experience on the one hand, and its geometric rationalization in modern cartographic discourse on the other. Moving from a consideration of sixteenth-century cosmographical thought (Chapter 1), to the practice of estate surveying in England (Chapter 2) and finally to the political geography of Ireland as imagined by English reformers (Chapter 3), my intention is to show how in all three of these seemingly unrelated contexts the ancient tradition of a sacred geometry – a kind of divine script reflecting the hidden truths of an invisible world – could be instrumentalized, across a wide spectrum of cartographic practice, for a variety of deeply pragmatic ends.

Historians of the sixteenth century have long recognized the importance of cartography as a means of controlling and making navigable the physical world, both within Europe and overseas:

> [F]rom the early sixteenth century, thanks to a new mathematical interest in cartographical projections that could take account of the curvature of the earth, more accurate assessments of the degrees of latitude, and to the challenge of ever-expanding knowledge of the world's surface, the cartography of Europe began to enable Europeans to imagine, believably, the geographical space in which they lived.[12]

The advent of a busy world of trade and commerce, and the rise of a culture of commodities enabled by the continuous exchange of material goods dependent on transglobal maritime trade routes,[13] would be unthinkable without the privileged access to physical space offered by cartography. But sixteenth-century maps were more than mere catalysts for economic growth:

> Landlords wanted estate maps, governments administrative ones for purposes of tax and toll control and the plotting of roads and canals, defensive fortifications and troop control assembly points. Statesmen used them for strategic purposes. Monarchs commissioned them as symbols of power. All over Europe they became part of the mental furniture of educated men [*sic*!]: indeed of their actual furniture, framed and hung, painted on walls, woven into tapestries, whole collections rolled or folded in chests and on shelves.[14]

Evidently, maps served a broad range of divergent interests: they are items at once highly practical, eminently political and overtly symbolic; and just as the early modern discourse of geography was both a scholarly pursuit and a commercial activity,[15] they are records of a significant increase in spatial knowledge as well as valuable economic objects in their own right.

Part II analyses the historically specific signifying power of the cartographic image by looking at various Elizabethan and Jacobean maps of Britain and Ireland. These maps, I aim to show, all make different statements about the territory they put on open display; in particular, they visually encode different perspectives on the territorial, political and conceptual struggles attendant on the emergence of a nation. Arguing first for the necessity of acknowledging the ambiguous status of maps in contemporary eyes (Chapter 4), I move on to examine the representational interplay of visibility and shadow in a series of maps in which a newly triumphant nation (Chapter 5) and its (potential) colonial extension in Ireland (Chapter 6), as well as the imagined relationship between both realms, are conceptualized in strikingly different ways. Early modern geographical discourse was neither fixed nor static but a highly flexible means of organizing cultural knowledge; and it enabled geographers to construct a range of different, even contradictory, graphic models for the political and social space of the nation. My discussion of cartography in these chapters is ultimately less motivated by an interest in the history of *maps* than in the culture of *mapping*. The distinction is an important one. Looking in detail at a historical sequence of single maps of a particular region leads to crucial insights into map production techniques and into the changing understanding of the extent and shape of a specific geographical space, but to focus instead on the process of *mapping* and its attendant cultural engagements is to regard the individual map, in Denis Cosgrove's words, 'as a hinge around which pivot whole systems of meaning, both prior and subsequent to its technical and mechanical production.'[16] It is this cartographic investment in the production of cultural meaning which is the principal focus of my analysis.

The rupture of scales in cosmographical thought considered in Part I draws attention to another dichotomy relevant to the contemporary geographical imagination, the theoretical distinction between the global and the local, or the whole and its parts. The rediscovery of Ptolemaic cosmography, although its technical 'backwardness' was soon only too apparent, paradoxically assisted what might be considered an early modern process of *globalization*: it

> offered to modern geographers a three-quarters empty canvas, leaving them free to inscribe on it the delineation of newly 'invented' or discovered lands; a form, at once closed and open, full and lacunary, that represented the ideal construction in which to house, with

their approximative and disparate localizations, the 'bits' of space that navigators brought back from their distant voyages.[17]

Parallel to these transoceanic quests the 'known' space of Europe was the object of an equally important movement of geographic 'discovery', one that explored not the global frame but the deeply historic and rigorously specific *locality* of place in an unprecedented wealth of regional detail. The importance of 'chorography' – in the eyes of many contemporaries, a new and exciting topographical interest – for a process of national 'self-discovery' in Britain is evident in the many regional descriptions coming from the presses in Elizabethan and Jacobean times.[18] Following Ptolemy, the difference between the local and the global was frequently rationalized as one between qualitative and quantitative modes of spatialization: chorography produced 'landscapes', in the aesthetic sense of the word, cosmography was predominantly – though not exclusively – a mathematical operation.

Yet to insist on the general applicability to chorographic works of the Ptolemaic distinction between quality and quantity is to ignore that the local could equally be the object of a largely quantitative rationalization, specifically if the spur to geographical activity was openly economic in kind. Thus, while my focus in this study is principally on the local – on the English and Irish 'part of the whole' – I endorse the theoretical premise that the kind of spatial transformations I intend to examine affected all levels of geographical thought, albeit to different degrees – a premise that leads me in the opening two chapters to look both at cosmography and at the discourse of estate surveying in England, embracing in turn the very emblems of the global and the local. The dominance of the latter in chorographical writing is my main focus in Part III which looks at the way spatial narratives translate cartographic uncertainties about the nation's external geopolitical frame into conflicting readings of the internal social configuration of the national territory. These readings, I argue, oscillate between the different models of social communities appropriate to the alternative textual forms of itinerary and map (Chapter 7). A more elaborate version of this distinction marks the gap between the respective spatial frameworks of Spenser's *Faerie Queene* and Drayton's *Poly-Olbion* (Chapter 8); and it is again instrumentalized, I finally argue, in the narrative inscription of Ireland with the discourses of nomadism and savagery (Chapter 9).

To say that in early modern geographical thought the expansionist drive was accompanied by an inward turn – a twin impulse of intellectual curiosity reflected elsewhere in the roughly parallel inventions of telescope and microscope – is to register the complex and multilayered nature of contemporary spatial consciousness, and hence to accord a significance to 'space' as an analytical category that already surpasses its narrow geographical sense. Space, as I will be using the term throughout, refers neither to an all-inclusive and infinitely transparent container of a physical world largely made up of material objects, nor to a realm of consciousness preceding all sensory phenomena: an empirical given best described by the natural sciences. Rather, drawing partly on the theory of social space developed by Henri Lefebvre,[19] I will consider space as the imaginative product of social (and political) action. Lefebvre argues for a theoretical analysis of the way space is not simply 'there', like an 'aggregate of sensory data [or] a void packed like a parcel with various contents',[20] but actively 'produced' – for instance, in literary discourse, but even more importantly within material practices such as architecture, town planning or civil engineering. In this study I start from the premise that geographical texts and images also 'produce' space in this sense, even though I am ultimately concerned less with the tangible and immediate effects of spatial orderings on actual social experience than with the signifying structures inscribed in the verbal and visual representation of differently imagined social and political spaces.

Can maps function as records of social space? If it is accepted that social relations *are* 'embedded in spatiality',[21] then the most recent definition of maps developed by historians of cartography implies an answer in the affirmative: 'Maps are graphic representations that facilitate a spatial understanding of things, concepts, conditions, processes, or events in the human world.'[22] The wide conceptual range of the phrase 'graphic representation', which does not reduce the map to a specific type of artefact, is possibly one of the advantages of this definition since it allows, among other things, an approach to the study of maps that may easily supplant the traditional Eurocentric bias of cartographic history. Yet to reduce maps entirely to their effects on the viewer and to disallow certain formal properties as constitutive marks of their generic affiliation is not a very practicable suggestion within the context of my project. John Gillies, in a recent critical evaluation of this attempt at a general definition across time and place, has commented that by identifying any kind of spatial image as a map 'the field of cartography is not so much expanded as exploded.'[23]

My own approach takes into account what contemporaries themselves would have considered maps (a point discussed in Chapter 4).

This proposition means to recognize that in early modern times, much as in our own, geographical knowledge as publicly exhibited in the practice of cartography embraced not merely topographical information in the narrow sense – the shapes of coastlines, the distribution of hills and valleys, the distances between human settlements – but attended to a whole cluster of issues we might be inclined to separate into different disciplinary compartments: maps could (and did) provide political, ethnological, strategic, social and linguistic information; they were instrumentalized either as ideal models in whose image an inadequate reality had to be fashioned, as purposeful interventions in contemporary political debates, or as reminders of the proximity of religious and ethnic difference; they dealt, in short, with multiple, overlapping and frequently contradictory forms of cultural and political identity. This fluid nature of cartographic semantics suggests that space – both in its narrowly geographical and its more broadly social sense – is always only provisionally conceptualized in maps and texts, that our representations of it are experimental in kind, and that verbal and visual 'acts of *mapping*'[24] warrant the sustained critical attention to their changing cultural signification.

Part I
Measurements

Introduction

When Christopher Marlowe's Tamburlaine stormed onto the London stage in the late 1580s, '[t]hreatening the world with high astounding terms, / [a]nd scourging kingdoms with his conquering sword',[1] his dream of a global empire both repelled and electrified Elizabethan audiences. For on the one hand, the barbaric Tamburlaine, Marlowe's theatrical version of a fourteenth-century Scythian shepherd, is an aggressive figure of dramatic hubris: epitomizing the danger of geographic and ethical transgression, his course of conquest and carnage through the whole of Asia is as spatially excessive as it is morally reprehensible. Yet not the least of Tamburlaine's redeeming features, in contemporary eyes, was his triumphant stage victory over the Turk: here was an awe-inspiring conqueror – closing ranks with the heroic champions of the ancient world, Alexander and Caesar – who in one grand military coup annihilated what persisted in communal European fantasy as the sabre-rattling, world-threatening Ottoman menace, an apocalyptic nightmare of cultural and religious terror. Current historical crises in Elizabethan England invite a translation of the imaginary Turkish threat into the very real fear of a Spanish invasion: 'we haue seene them on our coaste, and heard the thunder of their shot, in how cruell manner the proud disdaynfull insolent Spaniard of late daies hath threatned our Cuntrie, Queene and Citie, with fire and sword'.[2] This was the voice of a nation pushed to the edge of Europe, both geographically and politically – Tamburlaine's imperial aspirations, his extravagant desire for territorial extension, may well draw on a deep contemporary longing for an escape from a besieged insular existence.

Geography, as has long been noted, provides the action of the play with its forward thrust. The script, full of foreign place-names and references to exotic locations, reads like a veritable index of geographical diversity and wonder. Returning from a round trip through Africa, Tamburlaine's general Techelles reports on his progress:

Techelles   . . . I have march'd along the river Nile
              To Machda, where the mighty Christian priest,
              Call'd John the Great, sits in a milk-white robe,
              Whose triple mitre I did take by force,
              And made him swear obedience to my crown.
              From thence unto Cazates did I march
              Where Amazonians met me in the field,
              With whom, being women, I vouchsaf'd a league,
              And with my power did march on to Zanzibar,
              The western part of Afric, where I view'd
              The Ethiopian sea, rivers and lakes,
              But neither man nor child in all the land.
              Therefore I took my course to Manico,
              Where, unresisted, I remov'd my camp;
              And, by the coast of Byather, at last
              I came to Cubar, where the negroes dwell,
              And, conquering that, made haste to Nubia;
              There, having sack'd Borno, the kingly seat,
              I took the king and led him bound in chains
              Unto Damasco, where I stay'd before.

                            (2 *Tamb*, I, vi, 59–78)

Given the geographical excess of its imagery, *Tamburlaine* is perhaps best read as an imaginative dramatic testimony to the impact of cartography on sixteenth-century culture.[3] Marlowe's characters go on a rampage not around the real-life landscapes of Persia, Turkey or Egypt, but 'walk with their fingers' across the world maps coming off the presses of the Flemish cosmographers Ortelius and Mercator: every toponymic attribution in the report of Techelles' smooth African campaign – Prester John's Abyssinian empire, an Amazonian kingdom near Mozambique, a black Sahara, even a transplanted Zanzibar bordering on the *Oceanus Aethiopicus* – is based directly on the verbal and visual reference provided by the map of Africa included in Ortelius' world atlas (Plate 2).[4] In a very fundamental sense Ortelius' new map of the globe, first issued in 1570, is a defining structural principle of the entire play – more than merely its referential paradigm, it acts almost literally as the dramatic scene of action.[5]

One of the prime moments of the play's cartographic obsession occurs in Act IV as Tamburlaine lays siege to the city of Damascus. Zenocrate, an Egyptian princess captured in one of Tamburlaine's previous raids, pleads with him to refrain from attacking the city – the town of her father – and settle for a 'friendly truce'. Tamburlaine replies:

Tamburlaine   Zenocrate, were Egypt Jove's own land,
            Yet would I with my sword make Jove to stoop.
            I will confute those blind geographers
            That make a triple region in the world,
            Excluding regions which I mean to trace,
            And with this pen reduce them to a map,
            Calling the provinces, cities, and towns,
            After my name and thine, Zenocrate.
            Here at Damascus will I make the point
            That shall begin the perpendicular:
            And wouldst thou have me buy thy father's love,
            With such a loss? . . .

(1 *Tamb*, IV, iv, 79–90)

As Tamburlaine's sword transforms into a cartographer's compass the map turns into an imaginative correlate of his monstrous military exploits, the fantasy of an empty canvas on which to paint his own private empire. Generating new realities, a new toponymic world order, the map does not passively reflect but actively shapes the world of the play: Tamburlaine's dramatic regime, emphatically defined against the outdated products of 'blind geographers', is entirely dependent upon this violent act of cartographic creation. That Tamburlaine should deride the rejected image of the world – the 'triple region' – as no more than the pitiful display of a geographic *blindness* is a reminder of the power of vision associated with modern cartography: its apparent lucidity and far-reaching sight, its unrivalled capacity – as enthusiasts like Thomas Blundeville were quick to note – to make us *see*.[6] Geography, Ortelius wrote in the preface to his world atlas, is 'the *eye* of History';[7] that is to say, maps are not mere visual aids to memory, nor do they simply illustrate the spatial setting or colourful scenery for the drama of human affairs – rather, they are an enabling condition of historical action.

Two cosmographical paradigms were historically available that divided the human world into a 'triple region': the TO map commonly associated with Isidore of Seville; and the world map described in Ptolemy's *Geography*, possibly the most influential geographical text circulating in early modern Europe. On Isidore's map, the world appears as O-shaped, describing a perfect circle, which is internally divided by a T into the three continents of Asia, Europe and Africa. Jerusalem, the world's sacral centre, is the point where the two bars of the T meet. Ptolemy's map, on the other hand, looks forward to the new 'scientific' geography of sixteenth-century Europe by constructing a decentred projection of the globe that lacks a focus, an *omphalos*, and is exclusively generated by the application of mathematical knowledge to the physical world. But although it represents, due to its specific textual history, a later state of European geographical consciousness than the TO map, it equally prompts Tamburlaine's derision: locked in the classical paradigm of a human sphere structured around the 'middle of the earth', the *Mediterranean*, it locates the globe's only inhabited (in fact, according to the Ptolemaic episteme, its only *inhabitable*) region in the northern hemisphere – and thus, by ignoring the land masses both on the west coast of the Atlantic and south of the equator, equally reduces the world to the 'triple region' of Asia, Europe and Africa.

Ptolemy, therefore, joins the 'blind geographers' of the medieval *mappae mundi*: the passage denounces all manners of 'triple region' maps, whether generated by the sacred symbolism of the TO map or the spatial reduction of the Ptolemaic image. All are superseded by Tamburlaine's actions: rejecting traditional geographies, he intends to engender his own spatial representations. Yet, as John Gillies has noted,[8] his reference to a central 'point' is, strictly speaking, a theoretical absurdity on a map subjected to the mathematical abstraction of cartographic scale which denies representational hierarchies and reduces orientation in space to a mere referential function of the geometric grid. The placing of Damascus in a central 'point', whether intended to supplant Jerusalem or not, thus exposes the contradictions of Tamburlaine's own cartographic conceit, allowing us to get an ironic glimpse of the 'profound incongruity, the incommensurability, of the medieval and the Renaissance constructions of space'.[9]

A further contrast between two opposing and essentially incompatible mental maps is suggested by the immediate social context in which the reference occurs. Tamburlaine contemplates military action – he wants to conquer Damascus; Zenocrate foregrounds family ties – she wants to save the city of her father. But the need to 'confute those blind geographers' – perhaps in itself little more than a rebellion against paternal authority – clearly takes precedence over a sense of filial obligation. At least three different discursive layers shape the social nature of this confrontation. First, it is a deeply *gendered* encounter: a powerful man ignores the request of a powerless woman. Zenocrate is treated as little more than a spoil of war who, in the later course of the play, is eventually married off to Tamburlaine against the backdrop of a ruined Damascus. Military and sexual triumph coincide to confirm the dominance of patriarchal power. Second, the scene is marked by a discourse of *class*: Zenocrate is a princess, daughter of a king; Tamburlaine a lowly shepherd, whose rise to power mocks the 'natural' order of traditional social hierarchies. From this perspective, Tamburlaine is cast as the representative of a pastoral world gone horribly wrong, and his abuse of an oriental princess stages the fear of popular rebellion – he is as much a conquering tyrant as a kinsman of the home-grown Kentish rebel Jack Cade. Third – and this is the thread I will be pursuing in the following chapters – the scene gestures at an *erasure of social space*: Zenocrate pleads for a recognition of the social bond tying her to her father, Tamburlaine overrules her with his imperial aspirations – he will conquer the world, and reshape, as well as rename, the globe in his own image. He will not settle for less: 'And wouldst thou have me buy thy father's love, / With such a loss?' Both Zenocrate and Tamburlaine measure out the same spatial location, Damascus, in radically different conceptual terms: for Zenocrate the city is both product and origin of deep social and human ties, to be cherished and protected; for Tamburlaine the city needs to be encoded as a point on a map he wishes to own and redraw, an imaginary point at the intersection of arbitrary geometric lines running across the globe.

In so far as Tamburlaine's violent cartography annihilates the identification of the city as a space of human bodies, I suggest, this brief scene stages the explosion of the spatial paradigm that encoded space as an inherently social category: space is no longer represented, nor experienced, in terms of human corporeality – giving way to a phenomenal upsurge of interest in the ancient knowledge of geometry. Beyond the dramatic absorption of cartographic energy the scene is thus evidence of two opposing conceptions of space – one social, one geometric – overlaying the medieval/renaissance divide of its cartographic referents. This confrontation is observable across a range of texts implicated, irrespective of genre, in a redefinition of contemporary spatiality: space defined in terms of the people using it – space shaped in the image of, and in intimacy with, the human body – increasingly yields to its representation as a mathematical diagram, as an abstract grid pattern inscribed on a blank sheet of paper.

This conflict over the cultural meaning of space is the focus of Part I. Taking my cue from *Tamburlaine* I will examine its significance first in

the intellectual control exercised over the human sphere in cosmographical thought, then – in a move from the global to the local – in
the discourse of English estate surveying, and finally in the political
geography envisaged by the English colonial administration as an effective tool of reform in Ireland. These three paradigms – mental ownership of the world, the commodification of land in an agrarian sphere
and its quantification for the purpose of military surveillance – may
appear as separate and quite distinct moments of spatial appropriation,
beyond the possibility of comparison. Yet all three spatial constructs –
globe, field, colony – not only serve as a didactic testing ground for a
rhetoric of political order, proper rule and a well-managed state, they
are further, and more germane to my enquiry, objects of a measuring
operation that defines a distinctly modern sense of space. The application of geometry to lived cultural spheres is not merely evidence of
a desire for internal control translated into outward practice, it also
signals the homogenizing force of a representational act – with relevance (as I shall argue below) for a sense of cultural identity at large –
that necessitates both a code of authoritative knowledge and a powerful privileged agent: just as the mathematical textbook claims access to
absolute truth, cosmographer and surveyor profess a representational,
indeed mimetic, accuracy approaching divine perfection. The chapter
that immediately follows will deal with the first of my three paradigms,
the discourse of cosmography: in the light of my reading of *Tamburlaine*,
it considers the way the construction of the globe in cosmographical
works increasingly elides, across the sixteenth century, a definition of
global space in explicitly social terms.

1
Mathematics of the World

The description of the world is a traditional intellectual pursuit, and the changing representations of the human sphere depend as much on moral and cultural frameworks imposed on the act of global description as on the extent of geographical knowledge accumulated by specific societies. It could also be an inherently dangerous act, bound to be viewed with suspicion by the powerful and mighty. The fate of Metius Pomposien, reported by the Byzantine historian Jean Xiphilin, well illustrates the political significance attached to the map of the world in ancient times. The Roman emperor Domitian, fearing a rival claim to the throne, had Metius put to death in 91 AD although, as Xiphilin writes, 'he was not accused of any other crime than having kept in his cabinet a map of the terrestrial globe.'[1] Keeping the globe a secret may seem an inadequate, or even impossible, way of ensuring political control, yet it continues to have an irresistible attraction for military strategists: though later centuries have grown less suspicious of the inherent immorality of maps and globes, cartography's susceptibility to censorship has hardly abated in our own time.

In the sixteenth century the genre of cosmography was subject to extensive transition. Nine decades prior to Tamburlaine's exorbitant bid to redraft the map of the world Hartmann Schedel, a native of Nuremberg, put together a comprehensive description of the entire human sphere known in English as the *Nuremberg Chronicle*.[2] An elaborately ornamented, carefully produced volume running to over 600 pages, the chronicle is a vast 'compendium of history, geography and the wonders of the world as viewed from medieval Nuremberg'.[3] The book is simultaneously the last and the first of its kind. It brings to an end the tradition of late medieval world chronicles yet constitutes a new departure in the history of print: while its modest subject matter – the complete history of the world – is still strictly encoded in the worn patterns of biblical chronology, the history of its material production announces the arrival of capitalist speculation in the late medieval book trade.[4] Authorship has been rightly granted to Hartmann Schedel but the book is really, like all cosmographical work, the result of a collective enterprise: produced in 1493 by a syndicate of humanist scholars, printers, and illustrators, living not only in the same city but effectively in the same street, the risky venture of large-scale publication nearly caused their financial ruin when it returned only disappointing sales figures.[5]

The *Nuremberg Chronicle* stretches the limits of historical writing, and it is its geographical emphasis that makes it most relevant to my enquiry. The general division into individual chapters follows the logic of the seven ages of the world, a biblical timescale fleshed out with the traditional contents of a universal history. This chronological narrative is frequently interrupted by verbal and visual references to individual cities: a brief discussion of whatever seems noteworthy about a particular town is accompanied by a woodcut providing an ocular impression. These illustrations are never simply realistic depictions, though some of the city views incorporate authentic information.[6] Most are intended to convey an image of the city as a symbolic structure, focusing on what makes them typical, not specific and particular.[7] Many are wholly imaginary and may be employed as often as ten times throughout the book, creating bewildering, transnational identities between, for instance, Mainz and Naples, or Marseille, Trier and Nice (Figure 1.1).

As the chronicle progresses, more and more cities are introduced into the narrative at the moment of their supposed historical formation. Ranging freely over the map of the world this approach is moral and religious but still fairly unresponsive to national patterns. Schedel's spatial typology differentiates principally between city and non-city, between civic space as the crystallization of a civilizing process and the uncertain geography of an unformed and unhistoricized landscape. Graphically, the city views, whether authentic or imaginary, portray exaggerated constructions of generic civic space: depicted as if viewed from an imaginary point on the ground, the horizon is raised to crowd as many individual buildings as possible into a single image; palaces, towers and churches – dominant moments of architectural achievement – marginalize lesser objects; fortified walls circumscribe the city as a defensive bulwark. Schedel's cities are stone labyrinths, monstrous and overwhelming cityscapes, that dominate their natural locations and erase their links with the surrounding landscape. His twin emphasis on history and geography, on time and space, produces mutual dependencies: while a radically temporalized global space is subordinated to a bib-

Tryer

Figure 1.1 Hartmann Schedel: imaginary view of Marseille, Trier and Nice (1493).

lical time frame, history itself, the memory of past thought and action, becomes the condition of imagining the globe as a vast catalogue of cities, a universal space shaped by human effort.

Judging from its well documented print history – which includes a pirate pocket edition published in Augsburg[8] – the *Nuremberg Chronicle* was aimed at a wide audience, yet such projects could still arouse considerable mistrust. Schedel's immediate contemporary Sebastian Brant, for instance, could well have included him on his *Ship of Fools* which set sail in 1494, one year after the original Latin and German editions of the *Nuremberg Chronicle* appeared in print. Among the many characters that perform their mad feats on his vessel, Brant took on board a geographer, engaged in 'the folysshe descripcion and inquisicion of dyuers contrees and regyons': 'Who that is besy to mesure and compare / The heuyn and erth and all the worlde large / Describynge the clymatis and folke of euery place / He is a fole and hath a greuous charge'.[9]

Brant was deeply suspicious of cosmography's universal aspirations and used his new recruit – whose geographical knowledge, specifically his navigational skills, would have no doubt come in very useful on the ship's mad voyage – to promote the classical injunction *Nosce te ipsum* which accorded moral privilege to the value of self-knowledge:

> Ye people that labour the worlde to mesure
> Therby to knowe the regyons of the same
> Knowe firste your self, that knowlege is moste sure
> For certaynly it is rebuke and shame
> For man to labour, onely for a name
> To knowe the compasse of all the worlde wyde
> Nat knowynge hym selfe, nor howe he sholde hym gyde.[10]

Barclay's liberal English translation here retains the drift of Brant's original version: to mark as morally and spatially excessive the violation of the borders delimiting man's natural place in the world, and to denounce as misguided the vain ambition endemic in a geographical knowledge which focuses by definition on spaces exterior to your own being – an extravagant worldliness specifically enabled by the 'unsure science of vayne geometry'.[11] One significant subtext of this passage is the ancient idea of the body as microcosm, constructed in analogy to the larger macrocosm encompassing god's creation in its entirety[12] – an analogy that sets up explicit affinities between medicine and geography, between the doctor and the cosmographer, and between the anatomical treatise and the cartographic atlas.[13]

In embracing the entire surface of the habitable world, the *Nuremberg Chronicle* clearly transgresses the moral imperative of Brant's *Ship of Fools*. But it may be seen to recapture the spatial limitations posited on geographical activity through another, less obvious angle. Although the space encompassed is of a global dimension – the catalogue of cities covers the whole of the known extension of the earth – it is still in a very immediate sense tied to the human body: cities are included in the chronicle as the only relevant geographical sites worthy of note on account of both their aptness as symbols celebrating the achievements of civilized man and their status as human dwelling places. In terms of their thematic relevance, the city views thus emphasize the essentially social nature of global space: the globe is imagined precisely through the lived space of the city. The only other consistent series of illustrations in the chronicle are numerous portraits depicting the powerful and famous (Figure 1.2). These are equally symbolic images: 44 differ-

Das drit alter

Dieweil Sela der sun Jr̄ de noch ein kind was da gab
ine iudas nit der Thamar die des her vnd Onā weib
was gewesen. sunder er sendet sie ein witwen wider in irs
vaters haws heym. aber do Sela gewachsen wz besoʒgt
er ine zegebē dz er nit ertödt wurdt als sein brüder. also ver
stellet sich Thamar als ein gemains weib vñ sasse an dẽ weg
schaid vñ empfieng von iuda vñ gepare phares vñ zaram.
Vincencius in seynem geschichtbuch setzt hie vonn Asse
nech ein schöne histoꝛi. wie sie gar schön vñ erbeꝛ vnd
doch dabey stolz vnd hochfertig weꝛ vnd alle mañ versme
het. aber wiewol sie erstlich den Jpseph zu kaīne mañ wolt
yedoch als sie sein schön waißheyt vñ beschaidēheit mercket
do begeret sie sein gentzlich. doch wolt er nit verwilligen sie
ließ dañ vor ir abgöter. vñ wiewol sie sich darūmb betri
bet so wardt sie doch auß englischer vndrichtung gelaubig.
Rhodis die stat. von der die innsel Licie Rhodus heißt.
ist voꝛ Cristi gepurt. vijᶜ. xl. iar zu Joseph zeiten võ dẽ
Telchinieꝛn vnd Cariatieꝛn die durch Phoꝛoneum den könig
argiuoꝛum. voꝛlangst überwunden warn gepawet woꝛden
vnd ist vnder den innseln die man ciclades auß vꝛsachen dẽ
gelerten wissende nennet: den thenen die vom aufgang dʒ sun
nen daselbst hin komen. die allererst: darimm dañ (als Pompo
nius schꝛeibt) do der grund der stat gegraben wardt ein ro
sen knöpflein gefunden darnach die stat vnd innsel Rhodis
genāt woꝛde sey dañ nach kriechyschē gezüge ist rhodis souil
als ein rosen. Dise innsel hat in irm vmbkraiß. ijᶜ. mal achttail
einer meyl. Voꝛ andern wunderperlichen dingen was da
selbst ein sawl. lxx. elnpogen hoch die Lyndius ein iunger Li
sipi machet. Dise stat hat vil kriegs vñ zuletst võ den türckẽ
erliden. vnd ist doch alweg durch die ritter sant Johansen
oꝛdens beschirmbt vnnd geledigt woꝛden   Rhodis

Figure 1.2 Hartmann Schedel: heads and houses (1493).

ent woodcuts of crowned heads serve to depict 270 kings, 28 effigies do the same for the 226 popes referred to in the text.[14] The remaining sketches, including a range of saints, poets and philosophers of both sexes, also portray types, not individuals.

Though city views and portraits are rarely connected in any specific historical sense, they form a structural unity: cities and people are points of direct human contact in history; their visual juxtaposition produces the sense of intimacy which presents the geography of the chronicle as a distinct social configuration. Yet, arguably, this relationship of intimate proximity between city and people is ultimately neutralized by the cosmographical aspirations of the *Nuremberg Chronicle* as a whole, and it is precisely the vanity ascribed to the universality of this paradigm – the mediation of global space through human corporeality – which Brant's foolish world-measurer is explicitly charged with. Alexander Barclay's English translation of 1509 is clearest on this point: the geographer's colossal hubris – 'day and nyght infix[ing] all his thought' – generates the outrageous desire '[t]o haue *the hole worlde within his body* brought'.[15] Although the globe is linked to the body on the strength of a cosmological analogy it is vanity to imagine your own being as partaking of a global dimension, even if the attempt remains a purely mental exercise: the body of the cosmographer can absorb the 'hole worlde' only at the cost of fatally neglecting scrutiny of the self. Brant may wrap up this injunction in an outdated moral terminology yet the passage foreshadows a wider representational incongruity between body and globe in geographical discourse – an incongruity that allows us to conceptualize the visual transformations of the global image as emerging from the experience of a fundamental spatial disparity, a massive rupture of geographic scales.[16]

The degree to which the traditional equation between body and space – the conceptual centre of the micro/macrocosm analogy – began to collide with the contemporary perception of global space is obliquely registered in a well known illustration from Peter Apian's *Cosmographia*, published three decades after the first edition of Schedel, in 1524.[17] Apian is concerned with the abstract framework of mathematical and geographical order in which knowledge of the world is embedded: the first part of his treatise is a systematic attempt to classify and define cosmography's technical vocabulary while a second part presents an overview of continents, countries and the location of cities. A brief opening section discussing the difference between geography and chorography is placed alongside four illustrations showing a globe next to a head, and a city next to an eye and an ear (Figure 1.3). These images describe an

Geographia.   *Eius fimilitudo.*

CHOROGRAPHIA QVID.

Horographia autem (Vernero dicente) quæ & Topographia dicitur, partialia quædam loca feorfum & abfolutè confiderat, abfque eorum ad feinuicem, & ad vniuerfum telluris ambitum compatatione. Omnia fiquidém, ac ferè minima in eis contenta tradit & profe quitur. Velut portus, villas, populos, riuulornm quoque decurfus, & quęcunque alia illis finitima, vt funt ædificia, domus, turres, mœnia, &c. Finis verò eiufdem in effigienda partilius loci fimilitudine confummabitur : veluti fi pictor aliquis aurem tantùm aut oculum defignaret depingeretǫ;.

Chorographia.   *Eius fimilitudo.*

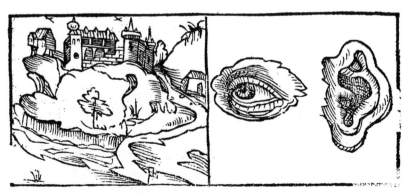

Figure 1.3 Peter Apian: the difference between geography and chorography (1524).

analogy between the art of painting and the discipline of geography: for a successful portrait a painter needs to start with the outline of the head in the same way that a geographer, wishing to describe the entire earth, needs to start with the circumference of the globe and the contours of continents.[18] The later addition of facial details – eyes, ears – corresponds to the work of chorography, or regional description, which concentrates on local topographies such as cities and rivers.

This definition is taken straight from Ptolemy and was repeated in near identical terms in almost every contemporary geographical textbook. But in spite of this ancient authority the comparison no longer worked, or rather, I suggest, it functioned only vertically, not horizontally. A city may be part of the globe in the same way that an eye is part of the head – both have their fixed location and function, both are central to our perception of space – but the globe, unlike the head, cannot actually be sighted (nor sited) as an object in space: its size far exceeds the visual capacity of the human eye. Apian's illustrations describe a fundamental incongruity: earth and head are graphically paralleled as if to suggest their essential similarity, their *identity*, yet beyond the text they drift off into different, and eventually into irreconcilable, spatial paradigms. The city can be 'lived' – as well as seen and heard, by eye and ear – in a very immediate sense as part of the spatial practice of the body; the globe can never be the object of a concrete sensual, or even visual, experience.[19] Just prior to this section Apian illustrates the meaning of cosmography with an image that projects the eye into one line with earth and universe, as conceptually related ideas, without acknowledging the fact that no human eye could actually ever see more than an insignificant subsection of the earth, let alone the universe, unless assisted by such visual aids as maps or globes. Apian's graphic interventions, I suggest, can be read as an ultimately unsuccessful bid to socialize the world, to invest it with a sense of corporeal intimacy, and to master and familiarize an expanding global construct requiring – as Apian's text, in fact, explains in great detail – a new representational technology that threatens to defeat his own prosopopoeic gesture.

The *Nuremberg Chronicle* is the product of a late medieval imagination. Conceiving of the world as a 'narrow, orderly place',[20] it testifies to a rhetoric of inclusion and identity which shuns novelty and explains the exotic in terms of the familiar. Published a year after Christopher Columbus had set foot in the New World, it did not register the intrusion of the practical men, the navigators and explorers, into the world of humanist learning.[21] The geographers of the sixteenth century, however, could no longer ignore the challenge posed by travel narratives and voyagers' tales to the authority of a scholarly world still in the thrall of a classical epistemology. Sebastian Münster's *Cosmographei, oder beschreibung aller länder*, 'Cosmography, or description of all countries', first published five decades after Schedel's chronicle, in 1544, needed to adopt a new, and different, spatial paradigm.[22] If Apian's *Cosmographia* represents one of two essential forms of cosmographical thought in the wake of the late medieval world chronicles – the mathematical treatise – Münster's *Cosmography* is the paradigmatic model of the alternative mode – the historical and descriptive work.[23] Eurocentrism, with regard to Münster, is not an accusation but an accurate description of the contents of the *Cosmography*: of six books, the first deals with general geographical matters – such as the history of creation, the forms and shapes of the natural world, or the division of the earth into continents; the next three, accounting for more than two-thirds of the book, are exclusively concerned with Europe; while it is only in the last two books that he turns to a discussion of Asia, Africa and the islands of the New World. A vast compendium of historical and ethnographic knowledge, the *Cosmography* proceeds on a logic of space: geography, not chronology, provides the pattern for a subdivision into chapters.

Münster opens his voluminous tome with a set of maps that delineate in succession first the world – both in its modern and ancient shape – then the traditional regions of Europe, and finally the continents of Asia, Africa and the New World – an order that roughly follows (except for the last item) the sequence laid down by Ptolemy. The immediate contrast with the *Nuremberg Chronicle* could hardly be stronger: Schedel included a single map on the very last page of his book – a crude outline of central Europe, or 'Germany' (in the Renaissance sense of the term). The first ever published map of that region, it sat uncomfortably at the end of the chronicle: added on together with other material too precious to be discarded (a few chorographic descriptions of European regions) it defined physical space in terms contrary to the rest of the book. For in selecting the city as the paradigmatic human place, Schedel chose the static, not the dynamic: although, as I have argued, his cities serve to conceive of the globe as a massive social space, they are not yet bustling centres of activity, vigorous sources of energy and life, but fully fixed and determined locations, defined by rigid social hierarchies. Münster's jumbled mix of city views and regional maps, on the other hand, pronounces a wholly different sense of physical space: the world is imagined in terms of spatial links, connections and networks, as a sphere of contact and communication.

The thumbnail sketches that preface each European section, like the odd perspectival view of Britain and Ireland (Figure 1.4) which presupposes a viewer at a point high above the English Channel, announce

Figure 1.4   Sebastian Münster: thumbnail sketch of Britain and Ireland (1544).

what emerges as a central characteristic of many of the city views scattered throughout the book: their relational, dynamic qualities. The plan of Trier may serve as an example (Figure 1.5): spread across an entire opening, Trier is featured in the centre of surrounding mountains, fields and villages; the view does not cut the city off from its context to celebrate its singularity, as was customary for the illustrations in Schedel, but emphasizes its relational location as part of a network reaching beyond the individual city. The majestic Moselle, prominently featured in the foreground, is not merely the town's water supply, its liquid lifeline, but also a dynamic trade route extending the city beyond its defensive walls. Trier is presented not as an item in a catalogue of place-names but as a town situated in contextual relation to other towns. Münster's theme, in other words, is connections across, not isolation in, physical space – a concern the *Cosmography* shares with the topographical map. Thus Münster's geography, his cartography in fact, describes a landscape of trade and travel, a space in which to move rather than rest, a spatial practice of incipient transglobal voyaging which ultimately marginalizes the isolated, fortified city – and the world atlas would soon reduce the city to little more than a mere point of reference on a geometric chart.

I should briefly stop at this point to consider the conceptual difference between representations of the city in maps or views and the regional or global map. It is obvious that the land depicted, say, on a world map outscores the city both by order of precedence and by its susceptibility to change. Physical landscape predates the arrival of man both in biblical and natural history and its essential forms – the hills and valleys, the

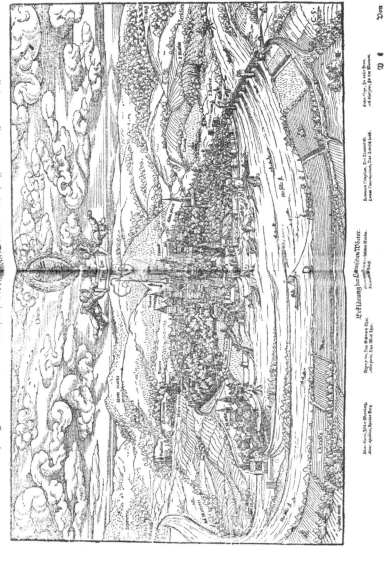

Figure 1.5   Sebastian Münster: view of Trier (1550).

rivers and coastlines – are invested with the quality of permanence; cities, however, are human inscriptions on the face of the earth. In the preface to his city atlas (discussed below), the Cologne printer Georg Braun presented this narrative of nature's precedence to culture in the shape of a gradual civilizing process, taking the form of a move from caves to huts and eventually to cities, or – in a parallel typology – from isolation in nature, to social bonding and human habitation.[24] In maps, the relative artificiality of cities introduces an element of mutability: as inescapably temporal images of human achievement they are subject to change, possibly quite radical change. The sixteenth century was familiar with more than enough classical precedents that illustrated how even the geographical location of civic space may not outlast the memory of man (obliquely recorded, perhaps, by the quibble in *The Tempest* about the correct site of Tunis). When cosmographical works began to reduce the city to a standardized cartographic symbol in the visual display of human geography, a mere point on the map – which they did from Ortelius and Mercator onwards[25] – they consciously opted for a mode of global representation that embraced the unchangeable, the timeless.

Perhaps it is wrong to say the city was marginalized in the tradition of the comprehensive geographical description, for in a sense it assumed a new importance by giving rise to its own distinctive genre. Georg Braun and Frans Hogenberg's *Civitates Orbis Terrarum* (1572–1617), 'The Towns of the World', is merely the best known and most comprehensive city atlas of many similar works that either preceded or were contemporaneous with this phenomenal showcase of an early modern metropolitan mentality.[26] In total, the *Civitates* collection includes depictions of more than 600 towns, spread across the whole range of the known extension of the globe. The first volume appeared in 1572, published simultaneously in Cologne and Antwerp, and featured 59 different plates showing a vast range of town plans and city views in a rich yet disordered display of contemporary civic consciousness. The compilation loosely follows the Ptolemaic itinerary: starting with Britain, it moves on to the Iberian peninsula to go clockwise round the Mediterranean, stopping off at the best known cities along the way. In his preface, Braun explained the overall aim of the collection:

> Simon Novellanus and Frans Hogenberg [the engravers] have, with highest skills and wonderful industry, shown how artful master builders have decorated *the entire earthly sphere* with cities and towns, bringing forth the shape of all cities in a manner so orderly, lively, and with respect to their proportions, that it seems you have before

your eyes not their imitation and counterfeit but the very cities themselves.[27]

The claim that looking at maps was the very equivalent of travelling to
the place depicted, not merely its feeble substitute, was repeated parrot-
fashion in the preface to nearly every sixteenth-century cartographic or
cosmographic work and served at least partly, I would suggest, to advertise maps as high-quality consumer goods.

My main interest in this passage is the dissonance between what is
explicitly stated as the cosmographic intentions of the collection and the
final shape it eventually assumed. Although the first book nominally
covers the whole of the globe (and was quite possibly intended to stand
on its own) it already shows a heavy bias towards Germany and Holland,
where the book was first published, and as subsequent volumes continue
to expand the collection and absorb new material at random, the project
increasingly suffers from a sense of internal fragmentation. The collection eventually ran to six volumes, published over a period of more than
four decades, and the pressure on the editors to include an ever larger
number of maps and views testifies, much against the original intention,
to the status of the city as the battle ground of the local, regional and
provincial. If the first volume produced an image of the globe dominated
by prestigious capitals such as London, Paris and Rome, Jerusalem, Con-
stantinople and Cairo, with Mexico and Cuzco documenting the most
recent acquisitions to a global sense of civic pride, later books in the series
needed to rest the earthly sphere on the feeble shoulders of towns like
Danish Husum, Westphalian Soest, Croatian Kostajnica or the island of
Chios in the Aegean Sea.[28] The more material the editors included, the
more they got caught up in ever more detailed chorographical descriptions at the expense of the cosmographical paradigm which initially
defined the project. The city, this development suggests, no longer served
to underwrite a script of world-embracing universality.

The individual representations adopt a variety of styles: some show
the surrounding countryside as a reservoir of manual labour and material goods, essential to meet the growing appetite of the city, some
display local and regional costume in the foreground, thus shaping
spatial particularity through the typical garments of the local population. Significantly, a number of images discard the conventional pictorial mode of representing the city from a human viewpoint in the
landscape and resort to the visual elevation of either the bird's eye view
or the embellished ground plan. According to Braun's own statistics in
the final volume, about half the images are traditional 'prospects', while

the other half constitute plans, depictions verging on scale-maps.[29] Strictly speaking, there are no purely 'geometric' maps at all in the *Civitates* compilation, and there is, in any case, still an important difference between the bird's eye view and the ground plan – the latter is unable to depict the size and splendour of a city's architecture, unashamedly reducing it to a mathematical patchwork of differently shaped plots on the ground (in frightening analogy, it should be noted, to the destructive force of a fire[30]). But both representational modes share the common feature of introducing into the pictorial set-up an imaginary and physically unattainable viewpoint, removed to divine heights, with the effect of greatly expanding the power of human eyesight. Of course, traditional views never respected the limitations of the human gaze: they frequently enlarged the field of vision, or stretched the laws of perspective, to include more on a single image than was strictly visible from any given spot. Schedel and Münster always privileged symbolic significance over 'realistic' features and in their city illustrations they were little bothered about vastly distorting architectural proportions. But on the plan or the bird's eye view something different happens: the former observes not perspectival but geometric proportion; the latter places the viewer in an identifiable yet physically unreachable location. Both representational modes thus consciously imitate an act of perception, but – significantly – an act that cannot be repeated in real life: the city opens up to its visual appropriation from a point of view conceptually removed from any possible spatial experience of the human body.

The *Civitates* collection is not the prime witness for this process – indeed, most people leafing through it will remember the images more for their proximity to landscape paintings than for their representational gestures in the direction of scale-maps. Yet the whole project emerged from the same milieu as the authoritative work that redefined the parameters of the discipline of geography in precisely these terms, Abraham Ortelius' *Theatrum Orbis Terrarum*, 'The Theatre of the Whole World', first published in 1570.[31] The alliance between the two projects is both personal and conceptual: Frans Hogenberg, for instance, one of the engravers of *Civitates Orbis Terrarum*, also cut copper plates for the *Theatrum*, and the congruence of Latin titles is an obvious indication of both works' generic companionship. Ortelius' *Theatrum* inaugurates the 'century of atlases':[32] though it was nominally Mercator who borrowed the name of a giant from Greek mythology to baptize his 1595 *Atlas sive Cosmographicae Meditationes*,[33] it was Ortelius' map collection which effectively gave birth to the genre. The first edition contained 53 different plates showing 70 maps, and throughout the four decades of its contemporary print history the maps more than doubled in number.[34]

There had been earlier map collections: portolan charts (navigational maps of coastlines) were frequently assembled in books of their own, and on specific demand from their customers sixteenth-century Italian mapsellers would bind individually printed maps together in a single volume. The Renaissance map room, such as the 'Sala delle Carte Geografiche' in the Florentine Palazzo Vecchio, is perhaps best thought of as a three-dimensional atlas. But never before had the whole world been systematically represented as a printed sequence of images with such formal coherence, or with a similar pretension to universality. Nor had any other cartographic project prior to the *Theatrum* enjoyed the same commercial success. With Ortelius' atlas the cosmographical project moved to an unprecedented level of spatial abstraction: the universal space of the earthly globe is encoded in a unified representational pattern based on a geometric projection which, in contrast to linear perspective, completely ignores the human observer – imagining a viewer in relation to the land depicted has become wholly irrelevant on the scale-map. The atlas both affords a totalizing overview and entertains the fiction of absolute spatial control: the world – imagined as a complete whole, as the sum total of all conceivable spatial relations – can be taken home to rest on a shelf, or decorate a wall.

But an attraction other than mere admiration for mathematical perfection accompanied the contemporary reception of these maps. The fascination of mental ownership engendered by these 'placeless' maps in the dislocated viewer is based on a paradoxical moment: while geographical diversity and cultural difference are erased by a fairly simplistic typology of cartographic signs, reducing the entire globe to a system of lines on a flat sheet of paper, their toponymic exoticism and suggestive pictorial shape trigger off a vertiginous geographic dynamism that is the very stuff of a cartographic poetics. In his remarkable study *Shakespeare and the Geography of Difference* John Gillies has identified the sensual, aesthetic qualities of the global image produced by Ortelius (Plate 3) as a '*semiosis* of desire': the absence of a secure, reassuring centre, the 'gravitational pull exerted by the two landmasses' east and west, the visual restlessness of the image, its 'constant discharge of energy'.[35] The Ortelian atlas conceives of the globe simultaneously as the object of a scientific and an erotic gaze; the maps are evidence of a bizarre congruence of the geometric and the mysterious, of the wish both to exert intellectual control over global space and to yield to the lure of the exotic.

This duality is in fact present in *Tamburlaine*, which makes such direct use of Ortelius' maps: the passages quoted above look at the world both as a mathematical construct ('Here at Damascus will I make the point/That shall begin the perpendicular') and as a space of danger and delight, resonating with unknown pleasures and erotic threats ('[I marched] unto Cazates . . ./Where Amazonians met me in the field'). The map I want to offer as a fitting emblem for this dual process is not taken from Ortelius' *Theatrum* but from later editions of Münster's *Cosmography* – an allegorical map of feminized Europe (Figure 1.6). This map was first introduced into the 1588 edition as part of an overall cartographic face-lift: the opening section of the *Cosmography* – containing a series of not very detailed scale-maps of the world, the continents and the countries of Europe – had remained virtually unchanged since the early editions (except for a number of maps added over the years). In the 1588 edition – the first for a decade – the publishers of the *Cosmography* acknowledged the aesthetic superiority of the *Theatrum* and based their introductory maps on the charts now provided by Ortelius. At the same time, a crude thumbnail sketch of Europe that had opened the second book of the *Cosmography* was replaced by the map depicting Europe in the shape of a woman.[36]

This simultaneous increase of accuracy and symbolism at a time when Ortelius' and Mercator's 'new geography' was the flavour of the moment suggests a paradigmatic change in the use of the space/body analogy that has dominated cosmographical discourse across the works I have been discussing. Most obviously, the body of the world is now clearly gendered: if Schedel's cities offered a social space of human bodies, irrespective of gender, a geographical space explicitly imagined as female assumes a host of qualities articulated by the signifier 'woman' in a patriarchal culture – passivity, fertility, penetrability, a need for male protection, a submissive return to the domestic. The act of sexualizing landscape in this fashion conceives of the global body as the object of an analytic masculine gaze; it does not encode the globe as a body-space suggesting a sense of corporeal intimacy. Yet the same book offers a second cartographic representation of the exact same geographical space: the Ortelian map in the opening section (Figure 1.7) which encodes the continent precisely on the premises of the male investigative gaze that generates the gendered space of feminized Europe. This cartographic partnership is evidence of the deep conceptual links between a dynamic male sphere of movement and travel, and a passive female space of inactivity and stasis.[37]

There is, of course, the possibility of reading the map of feminized Europe merely as an ironic comment on the *Cosmography's* earlier cartographic ignorance. Such a meaning may well have been intended but it hardly does critical justice to the richness of the symbolic act. Consider Münster's description of the inner workings of 'mother' earth:

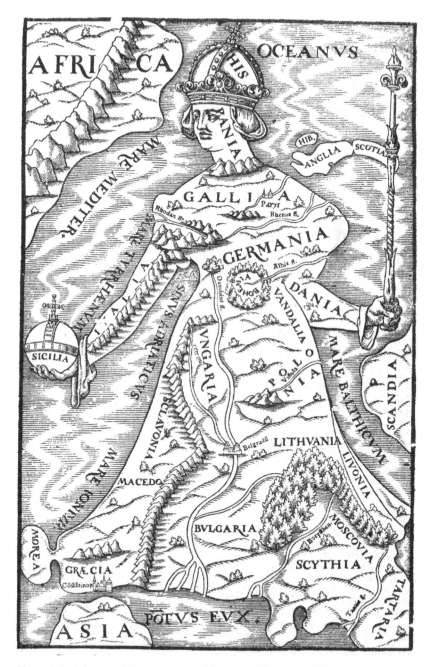

Figure 1.6   Sebastian Münster: map of feminized Europe (1588).

Figure 1.7   Sebastian Münster: map of Europe based on Ortelius (1588).

Just as heaven is god's dwelling-place, so is the earth the habitation, *even the mother*, of man and beast. Because it receives us the way we were born, nourishes and carries us while we're alive, and lastly receives us back into its bosom, and keeps our bodies till the day of reckoning.[38]

Further: earth is a 'friendly mother for man', she 'shows no anger like other elements', she 'gives birth to many peculiar things', 'serves every man', she offers us food and drink and paints the world in lively colours.[39] Feminized earth is a space of fertility and reproduction, a reservoir of colour and life, and a domestic haven required by the restless male voyager. But further on in the same passage earth is described as full of mysterious powers and dangers, affronting man with insatiable flames, poisonous vapours and destructive earthquakes.[40] Global space both requires and eludes masculine control; it willingly offers up its female body for male exploration yet threatens to forestall man's unrestricted progress. In translating these spatial complexities into different pictorial codes, the two maps of Europe emerge as intimately linked: the fantasy of mathematical accuracy and the conscious gesture of cartographic symbolism speak the same geographic language. The Ortelian conception of global space – geometric, quantifiable, a sphere of endless male voyaging – is the very condition of producing a symbolic map such as Münster's feminized Europe, a map that is not the antagonist of modern geography but a visual affirmation of its own erotic/exotic subtext.

The *Cosmography* was an immensely popular book and remained in print until 1628[41] – longer, in fact, than the *Theatrum*, though Ortelius' work was of course succeeded by the larger and more elaborate atlases of Mercator, Hondius, and the Visscher and Blaeu printing dynasties. I do not mean to say that Münster should be rated higher on the scale of geographical importance than Ortelius, merely suggest that the *Cosmography*'s lasting attraction may well be linked to its absorption of the modern Ortelian sense of global space. The *Theatrum*, however, does not contain anything as glaringly obvious as a map in the shape of a female body – be it of Europe, Africa or the New World. But the whole project, I want to maintain, partakes of the same conceptual thinking that provoked the map change in Münster. Ortelius' preface suggests as much: linking history to geography, as is customarily done in the period, he points to the problems a lack of geographical proficiency might cause in historical study. Without proper instruction in 'the knowledge of the countreys and places', historical narratives might be

cleane mistaken [by the reader] and otherwise vnderstood then they ought to bee: which thing commeth to passe in many discourses: but especially in the expeditions and voyages of great Kings, Captaines and Emperours: in the diuers and sundry shiftings of Nations from one place to another: and in the traueils and peregrinations of famous men, made into sundry countreys.[42]

The historical terrain charted here is one of male travel and exploration, of 'expeditions and voyages' suggestive of a dynamic, flexible sphere of ceaseless geographic expansion, shaped by the penetrating forays of travelling men into yielding, amorphous space. If Ortelius only allows an indirect glimpse at the gendered nature of his maps, consider this passage from the preface of a national atlas that was conceived in conscious imitation of the Ortelian model, John Speed's *Theatre of the Empire of Great Britain*, a regional atlas of Britain and Ireland, first published in 1611. Speed writes:

> The State of euery Kingdome well managed by prudent gouernment, seems to me to represent a Humane Body, guided by the soueraignty of the *Reasonable Soule*: the Country and Land it self representing the one, the Actions and state affaires the other. . . . And here [in Speed's atlas] first wee will (by Example of best Anatomists) propose to the view the *whole Body*, and *Monarchie* intire (as far as conueniently wee could comprise it) and after will dissect and lay open the particular Members, Veines and Ioints, (I meane the Shires, Riuers, Cities, and Townes) . . .[43]

The confrontation of body and soul, flesh and reason – translated without hesitation into the opposition between land and state – strikes me as far too resonant of a traditional male/female dichotomy to allow Speed's 'Humane Body' to remain without gender, specifically when the expressly marital rhetoric of the current English monarch, James I (to whom the atlas was dedicated), concerning the nature of the relationship between king and country is kept in mind. The reference to anatomy, to the 'culture of dissection' – in Jonathan Sawday's phrase[44] – is a reminder of the ruthless, clinical gaze that unites surgeons and geographers, that produces both body and space as objects of a modern 'scientific' quest pursued on the pages of two conceptually related books – the anatomical treatise and the cartographic atlas. Like the human body under the scrutiny of the anatomist, the interior space of the world is revealed, particularized and laid open, the cartographic drive accelerated to expose the innermost parts of the global body, its 'Members, Veines and Ioints'. For their successful conquests, both projects require an elaborate scientific arsenal of specified tools and a sophisticated technology of representation; both define our modern selves, in terms substantially different from a vanishing sense of social wholeness, as extensively researched and atomized bodies in fully mapped and compartmentalized geometric spaces.

A simplified summary would describe the conceptual development observable in the period covered by my three main examples – Schedel, Münster and Ortelius – as the move from catalogue to view to map; or from book to city to world; or, in terms more explicitly political, from hierarchical stasis to mercantile exchange to colonial expansion. I offer this concise typology fully realizing that none of these moments can be easily separated from each other. Nevertheless, cosmographical thought increasingly constructs the globe as a carefully measured world, a timeless geometric image. It subjects the cultural and geographic variety of the human sphere – so prevalent in Schedel and Münster – to the levelling and flattening sameness of mathematical order, and from this ground plan reconstructs the globe both as a sensual, erotic object and as an intelligible but – to recall my reading of *Tamburlaine* – ultimately a less socially responsive world.

Of course it would not be impossible to present the opposite narrative. Münster has clearly more interest in the 'social' than Schedel and the new Ortelian geography, precisely on the strength of its employment of a geometric representational framework, is perhaps the least obviously Eurocentric of all: the space of the *Theatrum* announces a degree of equality substantially different from the spatial hierarchies endemic in an earlier cosmographic tradition. Yet I would argue that this narrative, although entirely possible, ignores a number of significant developments: the slow, upward movement of the analytic gaze; the disembodied nature of a world subjected to cartographic scale; the discontinuities of our spatial experience on the level of actual practice; and the rupture between geographic scales that encode the world either as a space of lived social experience or as an object of a clinical, scientific quest. Beginning with deliberate artistry, with an image of the world enthralled to a kind of moral and social aesthetics, we arrive at a self-proclaimed new scientific world image, 'conscious of its novelty, confident of its superiority to the ancient geography'.[45] Schedel's image of the globe as a catalogue of cities and portraits, a universal space of heads and houses, shapes the world in the image of a human figure; the oceanic space embraced by the globalizing strategies of the new Ortelian geography moves away from this intimacy to colonize, as minutely and completely as possible, a landscape fully possessed only by the (masculine) mind.

2
Land Measuring: an Upstart Art

On the whole, the English contribution to the cosmographical project was negligible. The first independently authored book on the topic, though hardly an original contribution, appeared as late as 1559 when William Cuningham summarized standard continental knowledge in his *Cosmographical Glasse*, a mathematical treatise written in dialogue form.[1] Only fragments from Münster were ever translated,[2] an English edition of Ortelius' atlas did not reach print before 1606, and the first English version of Mercator was still many years off.[3] Maps, of course, hardly needed much translation, and atlases printed on the continent (in Latin and other languages) regularly flooded into England. Though English map-making was slow to gather momentum, the value of geometry as a generative force, as a universally applicable descriptive code, was readily acknowledged by English mathematicians. In 1588, London's newly elected mathematical lecturer Thomas Hood declared in his inaugural address:

> Let Geographie witnesse in vniuersall Mappes, let Topographie witnesse in seuerall Cardes, let Hydrographie witnesse in the Mariners plat, you your selues may witnesse in Martiall affaires, let the Gunner witnesse in planting his shot, witnesse the Surueior in measuring land, witnesse all those, that labor in mines, and those that practise conueying of waters, whose skill being tolde vs, we would scarsely beleeue it, were it not lying at our doores.[4]

That skill, Hood explained, was culled from Euclid, and his position is representative: geometry was never understood to be merely an arcane intellectual pursuit, of interest only to a select few, but a body of knowledge intended for immediate practical application. Robert Recorde, more trade-oriented than Hood, had trumpeted forty years earlier that 'numbre . . . [was] the ground of all menes affayres'.[5] A sharp awareness of substantial material changes affecting an increasingly mobile and dynamic world accompanied such contemporary eulogies on the value of geometry. Both across the globe and at home in England an ever-swelling tide of mathematical tracts, books describing new surveying technology or time-keeping devices, geometric manuals and geographic travel guides, and so on, swept across a mercantile space that needed to be planned and measured to make it economically viable and mentally navigable.

But in the 'vngratefull' and 'yron worlde' of mid-sixteenth-century England, Recorde feared in the introductory verses to *The Pathway to Knowledge*, general appreciation of geometry's usefulness was not as promptly forthcoming among merchants and artisans as it was among mathematicians. At least this is what the prolific mid-century author of several mathematical textbooks must have been thinking when he accused his intended readership – 'Merchauntes', 'Carpenters, Caruers, Joiners and Masons, / Painters and Limners', 'Broderers, Goldesmithes', 'Tailers and Shoomakers', 'weauers' and millers – of responding to his didactic efforts with 'sclaunderous reproch, and spitefull disdaine'.[6] Only a single professional group, Recorde's *persona* 'Geometry' laments, would show adequate respect: '[T]houghe other men vnthankfull will be, / Suruayers haue cause to make much of me.'[7] That land surveyors should naturally be thought to appreciate the need to acquire geometrical skills is not as self-evident an assumption as it might seem. In modern usage, surveying could be defined as the act of quantifying space along geometrical criteria in order to prepare statistical and cartographic records: a survey is a measuring operation that aims to determine the 'form and extent' of 'a tract of ground, coast-line, or any part of the earth's surface . . . so as to be able to delineate or describe it accurately and in detail' (OED, sense 5a). The Ortelian project of a comprehensive world atlas would have been inconceivable without significant advances in this area of applied geometry, and it is clearly to this meaning of the term that Recorde alluded.

It was not, however, the only meaning available. John Fitzherbert, author of the earliest printed English surveying manual, had something quite different in mind when he explained, in 1523, that '[t]he name of a surueyour is a frenche name, and is as moche to say in Englysshe as an ouerseer.'[8] What this 'overseer' was concerned with was not the shape of the land and its cartographic representation but a whole range of social and economic issues arising from the complex network of duties and responsibilities that defined the relationship between soil and subject, between tenant and lord. In Fitzherbert's tract, '[t]he description of a surveyor's duties revolves around administration of the court of survey, at which he was expected to examine records of tenure and receive tenants for their performance of homage and fealty';[9] what he was 'overseeing' was not land as the raw material of a map but land as a social space.

In this usage, surveying is synonymous with the general management of an estate, not with the measuring of land, and Fitzherbert's surveyor is probably better described as a land steward or an agrarian consultant.[10] It is indeed only in the sixteenth century that surveyors began to focus more on the inscription of boundary lines on the ground and their visualization in maps than on the social issues attendant on the running of an estate. Such changes signal the erosion of a conservative ideal which posited an agrarian world as governed by a moral imperative of mutual responsibilities, where the lord of the manor was still a paternal figure and not – as he was soon to become – an indifferent landowner. In other words, the conceptual change of the surveyor's function from steward to land measurer in the sixteenth century is a chapter in the larger story of the gradual shift from feudalism to capitalism, a process that saw land subjected to new economic forces

> which compounded demands for improved standards in the apportionment of property rights. Inflation and a rising population placed increased pressures on the land, while the dissolution of the monasteries stimulated the private property market by releasing hundreds of estates from the hand of the Church.[11]

The dynamics of a fluid land market affected the ways in which the whole practice of surveying was understood, and its main impact, in the latter part of the sixteenth century, was to gradually naturalize a perspective on agrarian space which foregrounded its status not as social realm but as marketable commodity.[12]

The high number of geometrical textbooks that either make extensive reference to new methods of surveying or are exclusively concerned with them is evidence that these changes did not escape contemporaries' notice. The surveyor's close association with more general patterns of agrarian change made him into a suspect figure, and many of the relevant manuals contain apologetic passages that set out to defend the new craft against all sorts of accusations levelled at it from outside, specifically – as we shall see in a moment – against accusations of inherent social bias. Thus, when a farmer in one of the most important of these textbooks, John Norden's *Surveyor's Dialogue*, asks a surveyor 'how such great persons [England's landlords] did before surueying came vp: for this is an vpstart arte found out of late, both measuring and plotting',[13] this ostensive naivety was an intentional authorial ploy to counter the accusation that the surveyor was merely a willing tool of rapacious landlords, '[going] like a Beare with a Chaine at his side'.[14] It allowed Norden – or rather his mouthpiece in the dialogue, the surveyor – to construct a narrative intended to demonstrate the long tradition of mathematical surveying which included as historical precedents the second chapter of *Zechariah*, where a man with a line sets out to measure Jerusalem, and the frequent inundation of the Nile Valley in antiquity, which necessitated measurements to re-identify the correct pattern of land ownership after the flood.[15] But these excursions into ancient wisdom were little more than feeble attempts to cover up 'the radical potential of the surveyors' activities'[16] in the present, by insisting on a respectable historical continuity that did not exist. As Thompson comments, 'the spur to [the early modern surveyors'] activities was not the joy of intellectual rediscovery but the gain to be made out of a changing economic and social situation.'[17]

Recorde was well aware that geometry might be perceived as a force of social polarization. Not only surveyors 'have cause to make much of' geometry, he continues the passage quoted above, 'so haue all Lordes, that landes do possesse'. Most significantly, surveyors and landlords join ranks against a third group: 'But Tennauntes I feare will like me the lesse. / Yet do I not wrong but measure all truely, / And yelde that full right to euerye man iustely.'[18] Mistrust of geometry, a sentiment evidently widespread among tenants, is immediately disqualified by reference to scientific impartiality; surveying clearly appeals to a higher order of truth and justice. Yet to suggest the mere possibility of its misuse is already to openly register geometry's potential complicity with forces of social division, even of political oppression – as Recorde's own conclusion implies: 'Proportion Geometricall hath no man *opprest*, / Yf anye bee wronged, I wishe it redrest'.[19] Other commentators adopted a similarly defensive stance. In a surveying tract by Edward Worsop, also written in dialogue form,[20] a clothier laments: 'The worlde was merier, before measurings were vsed then it hath beene since. A tenant in these daies must pay for euery foote, which is an extreme matter.'[21] This view, predictably, only prompts its immediate refutation by Worsop, who appears as a surveyor in his own text to explain that geometry is not a form of economic exploitation but a sign of England's spiritual health: 'True measure is not extremitie, but good iustice.'[22] The echo here of the ancient tradition of a sacred geometry, a divine script reflecting the hidden, unchanging truths of an invisible world, is distinctly at odds with the reality of an agrarian sphere where applied Euclidean knowledge helped establish a pattern not of truth but of covetous property relations – by facilitating, for instance, the practice of enclosures.

Norden's farmer runs through a whole litany of grievances. Surveyors, he rails, 'pry into mens tytles and estates . . . whereby [they] bring men and matter in question often times, that would (as long time they haue) lye without any question', they are 'the cords whereby poore men are drawne into seruitude and slauery', they cause 'fines [to be] inhaunced farre higher then euer before measuring of land and surueying came in'. Surveys may result in the loss of land, in the alteration of age-old customs and in extortionate rents.[23] Further on in the increasingly lop-sided dialogue, set against the backdrop of 'Beauland Manor' – a textbook abstraction of a real-life estate – the surveyor's superior rhetoric eventually moves the farmer to retract his earlier views. But the mere fact of the inclusion of such complaints in pamphlets intended to promote the 'art of surveying' indicates the anxieties generated by this popular reasoning and shows that the surveyor was liable to be perceived as a socially disruptive force. The advent of mathematical surveying, of 'measuring and plotting', marks the point where it is no longer possible 'to rely on local knowledge and an unbroken sequence of oral transmission to carry the record of property rights'.[24] The rise of new-style surveyors is indicative of a process that removed land from its location in popular memory and upset the tradition of a limited localized setting, where concepts of space could melt inseparably into the social sphere and 'the day's journey and the morning's ploughing'[25] were conventional units of measurement. What is at issue here is not only an increasingly instrumental understanding of the relationship between tenant and lord but a whole new conception of agrarian space.

This impact of surveying on the social world of early modern England is obliquely registered in the illustrations on the frontispiece of Aaron Rathborne's elaborate textbook *The Surveyor*,[26] published in 1616 (Figure 2.1). Two vignettes present the surveyor at work in the countryside. The lower image shows him in front of a widely used surveying instrument, a plane table, in the midst of a curiously depopulated landscape. Notable features of the image are a row of trees, a few hills in the background, a river and, most prominently, a castle in the top

Figure 2.1 Aaron Rathborne: frontispiece of *The Surveyor* (1616).

right-hand corner. The surveyor points at this castle with his left hand to indicate the main object of his survey. Unable to approach it physically, due to the intervening river, he will have to prove his skills by computing the distance to the castle purely with the aid of angular measurements. Behind the surveyor three figures are shown, assistants perhaps, or interested on-lookers. The surveyor, 'artifex', might just be in the midst of explaining his actions – we could be witnessing a lesson in practical surveying. The principal purpose of the landscape features – trees, hills, castle and the manor house in the top image – is to provide the surveyor with landmarks necessary for his geometrical operations: his gaze reduces them to fixed points visible through the sights of a ruler or a theodolite (the measuring instrument depicted in the top image). We are confronted, in other words, with a landscape abstracted from reality for the purposes of a survey. Both images are topological, not topographical in character – they portray a prearranged model landscape, ready for the intervention of the land-measurer.

The top image offers a suggestive allegory of the nature of this intervention. The surveyor is portrayed staring intently on a theodolite in front of him, with notebook in hand to keep a record of the measurements taken. As on the lower vignette, the landscape is encoded along the topology of a surveyor's *locus*, with one significant difference: the surveyor does not stand directly on the ground, but on top of two allegorical characters – 'a fool (in the cockscomb cap) and a faun (with pointed ears)'.[27] The general allegorical message seems obvious enough: the authority of science triumphs over the illusionary realm of fantasy, the truth of the theodolite marginalizes a world of myth and folklore.[28] Fool and faun are kept safely at bay by the surveyor (who appears utterly ignorant of the creatures underneath his feet) and the icon of Ortelius' 'new geography' – a globe lying on the ground next to the theodolite's tripod – remains untouched. The image, in short, champions the cool rationality of geometrical surveying which is actively engaged in suppressing the carnivalistic elements of a symbolic space inhabited by fool and faun – that is, allegorical representations of unreason and nature.[29] That Rathborne, on the frontispiece of a book almost exclusively concerned with the mathematical world of numbers and diagrams, should see the need to express this meaning by way of allegory is in itself an interesting paradox. But the image clearly demonstrates that the surveyor's elaborate instruments, such as theodolite or plane table, are hardly neutral icons of scientific progress. Rather, by enabling the translation of land into a set of tables and diagrams, such instruments, and the discourse of scientific 'truth' to which they belong, dispute the necessity of grounding land in the dense texture of a local and social script.[30]

The dynamics of this illustration can best be grasped, I suggest, with the help of Henri Lefebvre's theory of space. In Lefebvre's terms, surveying stages an intervention into the liminal, imaginative spaces of rural England, where fool and faun defined the symbolics of the land, to impose a centrally regulated and overtly coded 'representation of space' on the peasants' world. In *The Production of Space* Lefebvre argues for a theoretical analysis of the way space is 'produced', not merely in literary discourse but within material practices such as architecture or town planning. Setting up a conceptual triad, he distinguishes between:

- 'spatial practice' – the physical space that surrounds us: the house or the field, the city or the road, guiding the mechanical movement of everyday life and historically defining profile and spatial perception of a particular society;
- 'representations of space' – the mental, conceptualized space deliberately created by 'scientists, planners, urbanists, technocratic subdividers and social engineers ... [who] identify what is lived and perceived with what is conceived'; and
- 'representational spaces' – space experienced not as a given set of external parameters but as a flexible and imaginative category that defines our *social* existence, 'space as directly *lived* through its associated images and symbols, and hence the space of "inhabitants" and "users"'.[31]

Rathborne's frontispiece, I suggest, circumscribes a social space caught up between the latter two concepts: it imposes the dominant, public and overt forms of spatial configurations, tied to a visible and coded order of legible signs (the conceptualized space produced by the theodolite), on 'the dominated – and hence passively experienced – space which the imagination seeks to change and appropriate'[32] (the lived spaces of fool and faun).

This assault on the symbolic structure of 'unruly' social space was underwritten by a scheme of self-empowerment which is most closely associated, as the images on Rathborne's title-page suggest, with the devising of elaborate, geometry-based surveying instruments. The term geometry itself quite literally describes the measuring operation of the surveyor, as John Dee pointed out in his preface to the first English translation of Euclid: '*Geometria* ... is (according to the very etimologie of the word) Land measuring.'[33] It was Dee's friend and colleague, the mathematician Leonard Digges, whose geometrical works – the 1556 treatise *Tectonicon*,[34] an introductory textbook, and the longer and more sophisticated *Pantometria* of 1571,[35] published posthumously by his son Thomas – were among the first tracts to spell out in detail the immediate usefulness of Euclidean knowledge for the modern surveyor. Digges had learned his trade from a range of continental practitioners[36] and both his treatises were highly successful – the earlier *Tectonicon* was reprinted at least eight times before 1692.[37] His project, Digges thought, was revelatory in character: 'I am here prouoked *not to hide, but to open* . . . : yea, to publishe in this your tongue very shortlye, if God geue lyfe, a volume conteynynge the flowers of the Sciences Mathematicall'.[38] Laying something open which had been hidden before, disclosing – via geometry – truths otherwise unavailable: this precisely captures the contemporary fascination with the apparent novelty of geometrical insight.

In *Tectonicon*, a book astonishing in its graphic and conceptual simplicity, Digges describes a measuring instrument in revealing terms:

> [I]n my fantasy, the Instrument Geometricall, which is put forthe in thend of this booke, passeth al them and other [the outdated surveying equipment he has just described], for the exacte truth and quickest spede. This instrument . . . alone requireth a large boke, if it shoulde be sufficiently set forthe.[39]

This 'fantasy' of 'exacte truth' set forth in 'a large boke' – that is, truth equated with mathematical precision and mediated through the authority of the text – is occasioned by a comparatively simple device: the cross-staff, or 'profitable staff', derived from the astronomer's or Jacob's staff. Roche has commented on its impracticality in field survey, doubting even that 'such an instrument was ever constructed.'[40] But the object of Digges' praise is not so much the cross-staff itself as the principle of triangulation: knowing the degree of two angles and the length of one side in a triangle sufficed to compute the content and exact dimensions of the area covered.[41] This detachment from physical reality not only emphasized the sense of geometry's technical superiority; more than any other measuring technique it made possible early modern cartography. In Digges' *Pantometria*, a far more advanced measuring instrument makes its first appearance: the theodolite, a term of uncertain etymology, coined by Digges as a lasting contribution to the English language. For an arcane sixteenth-century expression the theodolite has had an impressive career, linguistically and technically, perhaps owing to the vague intellectual authority of its Greek sounding syllables. Radolph Agas praised it in the highest terms: 'it carrieth the forme of the first mouer, which commandeth all inferior creatures, and is preferred as most perfect and capeable, by the wisedome and ordinance of the Creator: so in vse and operation . . . this Theodolite commandeth euerie one of her subiects.'[42]

What concerns me here is less the importance of the theodolite and related instruments in the history of land measuring techniques – this has been analysed elsewhere[43] – but the ideological baggage attendant on their textual description which Agas' praise so vividly brings to the fore. In his view, to measure space with a theodolite is to emulate the *primum mobile*, the outermost sphere, origin of all motion. Further, as instrument turns into an instance of sovereignty the theodolite reigns supreme to attract the sacred aura of the divinely sanctioned ruler. A theodolite, for Agas, exercises power over form and subject. Two points deserve comment here. The first is to say that increasingly exotic surveying instruments were not only a response to practical problems encountered in the field but interventions in a struggle over the need for mathematical knowledge. In arguing this I am taking my cue from the historian of science J. A. Bennett who has identified the rise of the geometrically trained surveyor not merely as a sign of technical progress but as the result of a conscious 'propaganda effort' of Renaissance mathematicians, who realized that the traditional surveyor 'would need to be convinced of the value of a new type of surveying, requiring mathematical skills, and of a new image of the surveyor as a geometer, whose badge of office would not be a notebook and pole, but a theodelite.'[44] Instruments – and this is the second point – are thus implicated in the identity formation of the early modern surveyor whose professional existence was fashioned in the image of his superior technological competence. Thus Aaron Rathborne sounded a key theme in the discourse on surveying when he praised his 'absolute Instrument, which I call the Peractor, together with the making and vse of the Decimal Chayne, *vsed only by my selfe*'.[45] This last phrase neatly illustrates a dominant stance in the celebration of surveying equipment where the uniqueness of the instrument would define the uniqueness of its user. The land measurer was himself measured on the sophistication of his technology.[46]

At least one reason for this aggressive promotion of instruments can be glimpsed in the anxious prefaces, angry denunciations and the conspicuous use of certain pictorial forms characteristic of many surveying manuals. Norden's *Surveyor's Dialogue*, set on a fictional estate where characters slip in and out of roles, suggests that the spectacle of the surveyor in the field, parading his instruments and performing his measuring operation, surrounded – most likely – by rather suspicious farmers, had a distinct air of theatricality about it. Rathborne bitterly complained about dilettantes who look over a surveyor's shoulder just once before they 'prouide them[selves] of some cast Plaine Table, and within small time after, you shall heare them tell you *wonders*, and *what rare feats they can performe*'.[47] The same linguistic register, evoking the conscious mimicry of the stage, was used by Worsop: 'The ignorance of the time is such, that *to talke by roate* of measure, with *making show* of some instruments procureth great credit to such measurers.'[48] The illustrations that accompany some of the manuals gain particular significance when considered from this perspective. Digges' verbal explanations of surveying technology are graphically supported by depictions of instruments in active use. The figures handling them in the marginal images, perhaps in conscious allusion to the historical origins of geometry and mathematics, are usually clad in classical or oriental attire suggestive of theatrical costumes[49] (Figure 2.2). A treatise by Cyprian Lucar contains a fold-out that shows, like a visual prop list, the technical arsenal of the plane-table man (Figure 2.3).[50] Both vignettes on Rathborne's title-page feature a surveyor competently performing his role against the backdrop of stage-like scenery. Such illustrations may be principally intended to place the surveyor firmly in a contemporary public world but their effect was to undermine his bold, instrument-supported truth claims by implicitly drawing on the idiom of theatrical illusion.

Of course, the surveyor was more obviously 'making show' in a cartographic than a dramatic sense. If it is correct to say, as I have argued above, that surveyors produce frontal and overt spatial representations intended to erase the lived imaginative spaces of 'inhabitants' and 'users', then a direct articulation of this design is most likely to be found on the generic artefact his instruments helped produce, the estate map. The medieval surveyor was never in any sense a cartographer, but in the late sixteenth century land surveying increasingly became almost synonymous with map-making. The estate map has been defined as 'a plan of landed property, drawn not for a particular occasion or for some closely defined purpose' – like those medieval sketch maps used, for instance, in boundary disputes[51] – 'but for general reference'.[52] How are we to read these maps? Cartographic historians have provided two answers, one stressing the practical usefulness of such images, the other their ideological significance. The estate map, Harvey writes,

A BOOKE NAMED
Tectonicon,

Briefly ſhewing the exact meaſuring, and ſpe-
die reck̯oning all maner of *Land, Squares, Tim-*
ber, Stone, Steeples, Pillers, Globes, &c. Further, declaring the perfect making and large vſe of the Carpenters Ruler, con-
teining a quadrant *Geometricall: comprehending alſo the rare vſe of*
the Squire. And in the end a little Treatiſe adioyning, opening the com-
poſition and appliancie of an Inſtrument called the profitable
Staffe. With other things pleaſant and neceſſarie, moſt
conducible for Surueyers, Landmeaters,
Ioyners, Carpenters, and
Maſons.

Publiſhed *By* Leonard Digges *Gentleman, in the*
yeare of our Lord, 1556.

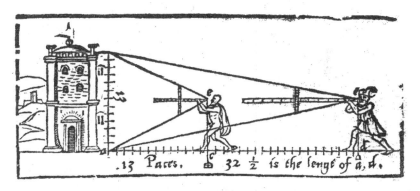

.13 Paces.   & 32 ½ is the lenge of a, d.

Imprinted at London by Thomas Orwin, dwelling
in Pater noſter Rowe, ouer agaynſt the   *S*
Checker. 1592.

Figure 2.2 Leonard Digges: title-page of *Tectonicon* (first published 1556).

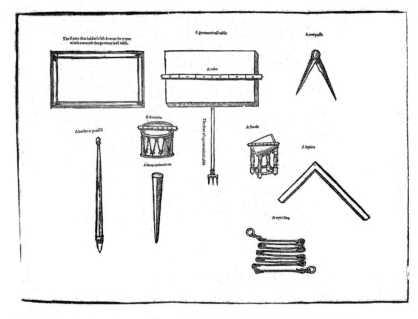

Figure 2.3   Cyprian Lucar: the arsenal of the plane table man (1590).

was a work the estate owner could consult for detailed information about the lands it showed; or he might point to it with pride, seeing it as a graphic epitome of his property, wealth, and social position. Often it was clearly designed for display, beautifully colored and elaborately ornamented. Often signs of wear, and many added corrections and annotations, testify to long service as a functional tool of estate management.[53]

So, on the one hand, estate maps offered privileged conceptual access to the land depicted; they 'would show [the landowner] at once, far better than the written survey, how his various pieces of land compared in size and position, and the precise effect of dividing or combining them for his own cultivation or for letting or for sale.'[54] Evidently, the forces of a nascent capitalist economic order, in which surveying was deeply implicated, have long been recognized as central reasons behind the rise of the estate map. At the same time such maps assumed extra-topographical, symbolic meaning. Their organizational principle was, of course, that all the land put on visual display formed part of one coherent manorial entity, 'the map was usually drawn to show nothing except the property of the landowner.'[55] Additional ornamental attributes, such as the frequent depiction of coats of arms, indicate that a further purpose of the cartographic image was to announce the lord's social status accruing to him from his belongings, and this overt exhibition of wealth and power is the key to its cultural significance. For the lord of the manor 'the map was one badge of his local authority.'[56]

The dual function of estate maps as both 'topographical inventories' and 'seigneurial emblem[s]',[57] is clearly visible on William Leybourn's model estate map (Figure 2.4).[58] The pictorial information contained in this map could be divided into two groups. The first supplies data on

Figure 2.4   William Leybourn: a model estate map (1654).

the land as such and on the uses to which it may be put: we are informed about the size of the estate, the distribution of fields, what purpose each plot serves. We also learn about the seat of power, the manor house, significantly placed directly in the centre of the image, at the point where all roads converge. The second set of data on the image defines the conditions of land ownership, both economic and conceptual: the title cartouche (top left) and the coat of arms (top right) claim the estate as the property of a particular lord; the 'scale of chains' in the bottom left-hand corner, and the compass rose in the right-hand margin, refer to the mode of exchange between image and world, the visual process of transmitting spatial information that constitutes the professional domain of the surveyor. The various connections between the different symbols on the map can be fairly complex; for instance, the title cartouche and the compass, simultaneously signs of authority and authorship, are linked through a relationship of patronage, and thus of economic dependence, between surveyor and lord of the manor.

The illustration thus appears to be a straightforward confirmation of the accepted reading of an estate map as both a practical tool and an icon of power. These two functions, however, do not just add up to an unproblematic duality of meaning but mask a representational tension which is not easily resolved – the map stages, in fact, a conflict over two rival conceptions of space. By organizing land according to the social indifference of geometric scale, the map indeed translates land into the idiom of the market: a commercial product that can be sold and exchanged, like any commodity. But its pictorial arrangement – the elaborate ornamentation, the stately manor house, the centrifugal visual effect – serves to present land as an extension of the lord's wealth and status, and thus asks the viewer to translate visual into social hierarchies. Yet this is no longer land that can be freely traded on the land market, for if it were, the map as a 'badge of authority' would disintegrate the minute the land it shows is implicated in an economic transaction. The cartographic construction of land as a geometric space, open to commercial speculation, would seem to constitute a potential threat to the integrity of the estate.

A similar faultline, separating what appear to be incompatible social and economic registers, runs through *The Surveyor's Dialogue*. On the one hand, Norden could describe the text as principally a catalogue of useful tools best suited 'to maintaine and lawfully to augment [the reuenewes of land]', and it is this focus on the increase of financial gain that necessitated accurate measurements to ensure 'that the manifest

truth might be confirmed, the hidden reuealed, and errors abandoned.'
Yet the idiom of value and profit is not the only textual perspective; it
competes with a residual feudal terminology that demands of the lord
and his tenants 'such a mutuall concurrence of loue and obedience in
the [tenant], and of ayd and protection in the [lord], as no hard measure
offered by the superiour, should make a iust breach of the loyaltie of
the inferiour'.[59] Maybe the text is indeed an attempt to evoke the imagined community of the estate in the face of rapid social change. More
likely, though, the separate social formations addressed in the split linguistic register of the dialogue are reflections of the precarious position
the surveyor occupies at the centre of two divergent sets of economic
interests. His professional interests must concur with those of his
employer, the lord, but it is the tenants who control access to accurate
information about the land they inhabit. Their cooperation is vital to
the success of the survey, hence the self-conscious liminality of the sur-
veyor-figure as the inquisitive agent mediating between the social
spheres of tenant and lord.

A survey, such passages also suggest, is a form of social control, and
it might be this function that at least partially resolves the representational tension characteristic of the estate map. Such maps, according to
Norden, are visual tools 'which tenants mislike, not that the thing it
selfe [the map] offendeth them, but that by it they are often preuented
or discouered of deceitfull purposes.' The lack of proper maps has
been the cause of 'infinite concealements', 'many intrusions and
incrochments'.[60] The tenants' names, Agas writes, 'shal be registred and
quoted in the same margent [of the book of survey] . . . whereby all con-
cealments and other abuses (if any shal be practised) shall not onely at
al times hereafter plainly appeare, but also readyly and truly helped and
reformed'.[61] Detection of fraud had been a traditional objective of the
'old-style' surveyor. The qualitative change, I would suggest, is the
increasingly desocialized nature of agrarian space. Map and attendant
register are envisaged as instruments of social supervision, forming – in
Norden's terms – 'a perpetuall glasse, wherein the estates of all the par-
ticulars within this Mannor, may be at all times seene and confirmed'.[62]
The names of individual tenants are little more than marginal ciphers,
ready to be erased at each contractual rearrangement. Reminiscent of
the standardized human figures that populate the cartographic margins
in the pictorial wall maps and atlases produced in the wake of Ortelius'
Theatrum,[63] the conversion of farmland into abstract geometric space on
the estate map, and the literal pushing of people to its representational
edge, does not so much challenge the lord's privilege as reinforce the

map owner's social authority and disciplinary power. People, as well as land, are 'scaled down' to come into proper view, a geometric map effect that affirms, rather than threatens, the validity of traditional social hierarchies.

This disciplinary aspect is already contained in the very word 'survey' with its revealing cognate 'surveillance'. The Foucauldian overtones of this semantic affinity strike me as entirely appropriate. 'Is not the eye surueyor for the whole body outward', Norden's surveyor asks, rhetorically conflating state and estate:

> And hath not euery common wealth ouerseers of like nature, which importeth as much as Surueyors? And is not euery Mannor a little common wealth, whereof the Tenants are the members, the Land the bulke, and the Lord the head? And doth it not follow, that this head should haue an ouerseer or Surueyor of the state and gouernment of the whole body?[64]

Gesturing at the fragile construction of both territorial and corporal boundaries, the initial body analogy circumscribes a suggestive ambiguity, casting the body both as object of, and agent for, the activity of the surveying eye. In both cases the surveyor inspects the body-terrain with definite visual authority, and in the sense that the head 'employs' the eye as its personal surveyor of the exterior world the image emphasizes the implicit congruence between the actual surveyor's view and the mental image this view impresses on the mind.

In the series of rhetorical questions that follows, body, manor and commonwealth are cast as interchangeable entities with the structure of each contained in the other. As a commonplace of contemporary political rhetoric this triple analogy is hardly a very striking move. Equally commonplace is the almost natural assumption that the commonwealth *is* in fact a body, with 'members', 'bulk' and 'head'. But unlike eye and state official, whose functions relative to body and commonwealth are of a permanent and indispensable nature, the surveyor – as Norden explains at great length in his book – constitutes an outsider to the estate he is called in to survey: his technical knowledge may facilitate its adequate management but is not essential to its everyday running. In placing the surveyor at the liminal point of contact between individual and commonwealth, which reflects his intermediary position between farmer and lord on the estate, the suggestion here is that the geometric conception of space his technology engenders is in need of incessant watch and control. In that his task resembles, is in fact metaphorically the same as, that of the state magistrates, the 'ouerseers' of the 'common wealth'. Ultimately, the surveyor's 'view' is identical with the perspective of power, imprinting official images of authority in the mind of the individual. Structures of state, not the immanence of a social world, were inscribed in the landscapes produced by the early modern surveyor.

Thus, if we read Norden's fictional farmer not simply as a rhetorical device intended to refute an oppositional voice – a strategy common to the dialogue as a literary genre – we may discover in his anti-surveying attitude more than a stubborn unwillingness to accept cultural and economic change. Instead, we find traces of a resistance against an increasingly desocialized conception of agrarian space, and against a cartographic practice which aligns the view through the sights of a theodolite with the perspective of a ruling elite. This opposition could forcefully articulate itself in the early modern period – in the Midlands enclosure riots, for instance, which took place in the same year that saw the publication of Norden's book – but on the pages of the *Surveyor's Dialogue*, where the measuring techniques central to the practice of enclosures were rehearsed in detail, the farmer's voice is hardly audible, absorbed as it is into a text that ultimately serves to erase all signs of such resistance: Norden's surveyor eventually proves so convincing that his erstwhile antagonist, the farmer, can even imagine taking up the land measurer's trade himself.[65] In subjecting a sense of social wholeness to the cartographic fantasy of abstract and homogeneous geometric space, the discourse on estate surveying – like the violent cartography of Tamburlaine – offers to efface rather than strengthen the deep conceptual links between people and land.

The literal centrality of the 'surveillant' lord is finally affirmed by Norden's sedentary image of 'the Lord sitting in his chayre' in front of the map, suggesting the relaxed ease of a wealthy landowner happy to 'see what he hath, where and how it lyeth, and in whose vse and occupation euery particular is *vpon the suddaine view*'.[66] The map, this passage shows, is evidence of the 'surveyor's controlling concern with defining property rights'[67] but also attests to his implication in techniques of supervision; he is instrumental in generating the fantasy of instantaneous social control identified with the calm contemplation of the estate map, an act of surveying repeated beyond the fields at domestic level. The cartographic synthesis of encoding the familiar in the universal thus serves to remove land from a social sphere where an uncharted locality still dominates the regional topography, placing it firmly in a world constructed by '[t]he perfect Science of Lines, Plaines, and Solides' which offers, according to John Dee, 'like a Diuine Iusticier, . . . vnto euery man, his owne.'[68] This rhetoric of ownership fostered by geometry, made manifest in the subjection of lived imaginative spaces of local 'users' to the abstraction of the scale-map, parallels the move in contemporary writings on Ireland where the idea of spatial homogenization through surveying is envisaged as a tool of reform to offer Irish space to *its* ostensible owner, the English monarch.

3
Surveying Ireland

In 1610 William Folkingham announced on the title-page of his *Fevdi-graphia* that modern surveying techniques, of such usefulness in England, were 'no lesse remarkable for all Vnder-takers in the Plantation of Ireland or Virginia'.[1] In setting the colonial realms of Ireland and Virginia in explicit relation to England this formula proposes to export, I want to argue in this chapter, not just efficient methods of manorial management but the structures of surveillance intrinsic to mathematical surveying. The desire to subject the Irish landscape to systematic surveying concurs with its frequent representation as 'a hostile wilderness, alien to civilized understanding, in which savage and implacable enemies fleetingly appear and disappear.'[2] In Ireland, the defining principle of landscape was not the immediacy of a rural world but the barbarous rebel who mistreated and wrongfully tyrannized Irish soil. In 1596 Spenser pitied 'the swete lande to be subiecte to soe manye evills' (577)[3], and in John Derricke's *Image of Ireland*, published in 1581, the dominant point of topographical reference is a 'pleasaunt lande' that has been savagely 'deformed through, / the life of Irishe karne'. The innocent land is forced to shelter the warlike and treacherous Irish foot-soldier, a 'monster' and a 'rebel', dangerous to the commonwealth, ignorant of religion, 'prone to sinfull lust'. In Derricke's diatribe the kern is rationalized as the transgressive inversion of the norms of civilization: 'Was neuer beast more brutishe, like / lesse voide of soueraignes feare. / No men so bare of heavenly grace, / more foes to Countries soile: / Nor traitours that doe more reioyce, / when thei their neighbours spoil.' The corrupt relationship the feminized land is forced to entertain with the Irish kern, foe to the 'Bride [that] is the Soile',[4] merges into the disturbing vision of a pre-cartographic landscape, subject to wild and 'barbaric' natives, at odds with the well-organized, administrative space produced by modern surveyors.

61

Folkingham's tract suggests that the links between English estate surveying and its practice in Ireland might be quite immediate and material. My principal aim in this chapter, however, is not to examine its
history in Ireland[5] but to trace the conceptual and rhetorical transfer
of the surveyor's analytic gaze from Norden's model estate 'Beauland
Manor' to the contested ground of Irish political geography. A brief historical excursion will clarify my approach. On 6 July 1567 Elizabeth sent
a letter to Sir Henry Sidney, Lord Deputy of Ireland, containing precise
instructions on her future Irish policies in the light of recent news about
'the late death of the troublesome rebelle Shane Oneile.'[6] Only two
decades later this letter had already been absorbed into the Irish section
of *Holinshed's Chronicles* which John Hooker brought up to date for the
second edition of 1587. This textual transmission in the service of the
construction of a national history may in itself not be a particularly
noteworthy event. What deserves closer attention, for the purpose of
this chapter, is the passage that immediately follows the mention of this
letter in *Holinshed's Chronicles*. Here Hooker notes how in the same year
the English surveyor and cartographer Robert Lythe, otherwise known
only for some maps of Calais he had drawn a decade earlier,[7] was dis-
patched to Ireland in the wake of this military success in Ulster:

> The queenes maiestie being deliuered from this traitorous rebell
> [Shane O'Neill], and hauing all Ulster at hir commandement and
> disposition, was verie desirous to haue a true plot of the whole land,
> wherby she might in some sort see the same, & did send ouer into
> Ireland one Robert Leeth [Lythe], skilfull in that art, and that he
> should make the perfect descriptions of the same.[8]

What concerns me here is the spatial logic that leads Hooker to link
Shane's violent death so prominently to the arrival in Ireland of a
modern map-maker whose surveys – through the use John Speed made
of them much later in his *Theatre of the Empire of Great Britain*[9] – 'eventually provided several generations of British and foreign map-users
with their cartographic image of central and southern Ireland.'[10]

Hooker's juxtaposition of the Irish rebel and the English map-maker
– one man's exit announcing another man's entrance – sheds light on
the conceptual role the topographical map had acquired in the Irish
context. At once a reminder of cartography's military rationale and of
the symbolic function of the queen's body, now 'deliuered from this
traitorous rebell', the passage charts a number of themes highlighted by
some recent analyses of English efforts to put early modern Ireland on

the map:[11] Hooker's use of the word 'description' implicitly defines the map as a means of 'gaining control over the world'[12] by systematically amassing topographical data which could be subjected to, and contained by, the cartographic grid of the flat surface, thus turning Ireland into a visible object of knowledge; the reference to the queen's desire to 'see' the landscape of Ulster 'in some sort' identifies the map as a visual record that could reveal a view otherwise invisible to the naked eye;[13] the sense that mapping constituted a form of political and ideological appropriation is apparent in the way Lythe's maps were to show how 'all Ulster' was 'at [the queen's] commandement and disposition'[14] – surely not a view all contemporaries would have shared; and the phrase 'true plot', spanning semantically both image and text, subtly subsumes the desire for political reform under the dynamics of cartographic representation.[15] The textual replacement of the rebel by the map-maker appears not as a historical accident but as the carefully constructed confrontation between two competing claims to the land. While the map will – or is at least expected to – create the image of a rebel-free landscape,[16] a plane space subject to the imperial gaze of the queen, pre-cartographic reality, by contrast, has the rebel roaming the wild, unconquered landscape at will, escaping the grip of culture and the fixity of the cartographer's plot alike.[17]

The brief passage from Hooker's Irish chronicle draws attention to the conceptual moment prior to cartographic production where the map's original subject matter – the land – is not conceived of as a straightforward empirical and geographical fact but as a site of conflicting social and political inscriptions. The abstract result of a survey – the geometric outline – requires the cartographer to move beyond a sense of land as a local and social space, deeply immersed in regional custom, that defies its translation into a set of mathematical data. This discursive model, rather than standing in conscious opposition to the land presented and contained on a map, frequently defines the moment of corruption that necessitates the reforming power of cartographic order. Thus, the colonial rhetoric surrounding the representation of Ireland constructs Irish space as the inherently transgressive realm of the savage or rebel where renewal of political control must be preceded by systematic cartographic description. As we have seen, the mechanism of this discursive logic was not exclusively restricted to Ireland. Estate surveyors in England, whose object of reform was not Irish alterity but the world of English peasantry, were equally engaged in the transformation and disruption of the locality of space, a conceptual parallel that suggests a degree of interaction, rather than a clear-cut separation, between the cultural spheres of England and Ireland. Both in the surveying manuals discussed above and in contemporary accounts of Ireland (to which the present chapter turns) the paradigmatic act of surveying heralds a mode of representing land which subsumes the social under the geometric and effaces attention to human detail by relying on the levelling impact of cartographic scale. In Ireland, the roles of cartographer and colonizer subtly melt into each other – a structural similarity that does not imply conceptual identity but draws attention to the continuous exchange between various discursive models as they are exported and re-imported in both directions across the Irish Sea.

Irish space, or rather space signified as 'Irish', is the object of specific rhetorical efforts in Edmund Spenser's dialogue treatise *A View of the Present State of Ireland* (1596). In the opening lines Irenius – the speaker in the dialogue in possession of local knowledge, fashioning himself as an expert on Ireland – speculates about the Janus-faced nature of the country: despite its 'goodlie and Comodious . . . soyle' (5) Ireland might turn out to be a 'secrete skourge' destined to haunt peaceful England. Irenius specifically constructs the island as a spatial threat in evoking the '*very Genius* of the soile' (12) as the presumable cause for both the present miserable state of the country and the potential dangers it might have in store for England. In a recurrent image, Irish-inhabited 'deserts and mountains' are pitted against the 'wide, open, and plain country' of the English invaders, a discursive contrast that serves as an explanatory framework for the current state of Irish lawlessness and English difficulties in maintaining military control over Ireland. Irish history after the conquest, Irenius admits, has been one lasting territorial struggle marked by the constant, unrestrained movement between opposing spatial paradigms. With the ground 'possessed and repossessed',[18] boundaries between culturally demarcated areas are never fixed but perpetually in motion, a process that attributes to the native Irish a kind of transgressive cartographic power and reduces land to the shifting coordinates of a battlefield, the mere prey of warring armies.

In official correspondence requests for new Irish maps, intended to counter such transgressive restructuring of space, are a frequent occurrence. The queen's letter of 6 July instructed Sidney to have certain areas in the north newly 'surveyde and discribid',[19] though it does not, contrary to the impression given by Hooker, mention the name of Lythe.[20] In an earlier letter Elizabeth had written that 'if you [Sidney] can fynde some skilfull person there that can make a more particular description [of Ulster], then already we have by any cart, it wolde helpe us to the under-standing of that which you shall wryte and satisfy us for all other conferences thereupon.'²¹ Clearly, the notion that maps could usefully assist the verbal description of place had become quite a commonplace assumption in government circles.²² This letter is also noteworthy because it predates by almost thirty years an analogous and recently much quoted passage in Spenser's *View* where the second speaker, Eudoxus, produces a map about half way through the text, declaring: 'I will take the mapp of Irelande before me and make myne eyes in the meane while my Scollemasters to guide my vnderstandinge to iudge of your plott' (3083–6).

With the help of the map Irenius, the dominant textual voice, proceeds to illustrate his projected policy of starving to death the most rebellious of the native Irish by placing four garrisons, each 2000 men strong, at strategic locations in Ulster with the aim to encircle and thus cut off the north-western section of the province.²³ Irenius' choice of north-west Ulster is suggestive, not as a conscious military strategy, but as an indication of cartography's own investment in 'undefined' space, for the topographically little known corner of Tyrone, Fermanagh and Tyrconnell constituted the most speculative area on all sixteenth-century maps of Ireland.²⁴ Textually, the map comes into play at the moment a direct encounter with the cartographically elusive 'flyinge enemye' (3062) is envisaged, leading Julia Lupton to suggest that the passage 'flags a series of connections between cartography and conquest which illuminates not only the political *content* or argument of Spenser's tract but also its *mode of discourse*, its enabling fantasies of political representation.'²⁵ Viewed as an illustration not just of Irenius' military policy but of his entire 'plot' the map promises a fully 'Englished' Ireland, a political vision projected forward in time by the text and grounded in space by the cartographic image.

One critic has argued that Eudoxus' map is a 'tool of empire'²⁶ which forces a unified spatial perspective upon the speakers in the dialogue and requires Irenius to adopt a more explicit and less metaphorical language. Thus, the elusive polyvocality of his earlier rhetoric, which confused the reader by constructing multiple symbolic versions of Ireland, is dropped because the visual force of the cartographic image disallows such textual vagueness. This analysis of the relative representational dynamics of maps and texts highlights the significance of this passage in the structure of the dialogue but presumes a stability in the relationship between cartographic sign and referent which is confirmed neither by the variety of early modern cartography, nor by the reference to the map in the *View*. What makes the map such a crucial discursive figure in Irenius' 'plot', in my view, is its provisional character, its power to enable an 'experiment' on Ireland's political geography,[27] rather than its status as a definite description. The textual adoption of a cartographic logic at this point in the dialogue enables its continuation precisely by opening up new perspectives, rather than by closing them down.

Put differently, in allowing Irenius to project his political fantasies into the future, the map introduces a relational perspective on Irish space which replaces the isolated internal view of the text's earlier sections. In analogy to the shifts we have observed in sixteenth-century cosmographical works – where Schedel's fortified cities first opened up to the surrounding landscape in Münster, and were then reduced by Ortelius to a strategic network of points assembled on a plane surface – space here becomes accessible by being first divested of its impenetrable boundaries and then subsumed under the uniform grid pattern of the flexible military chart. Before Irenius advances his cartographically inspired proposal of a military solution, in his analysis of Ireland's recent history and the complexities of its culture, he takes textual possession of Irish landscape by subjecting it to a range of discursive opposites: he points to the discrepancy between the land's agricultural potential and its actual miserable state due to Irish mismanagement; he presents the country either as empty, uninhabited and 'virginal', or as hopelessly ruined by its native inhabitants; etymologically and historically the land is appropriated as 'part of an ancient *British* unity' while the people are cast 'in the role of the intractable [*Scythian*] "otherness" ';[28] Irish inhabited 'desertes and mountaynes' (386), as noted above, are conceptually distinct from the 'wide Countrie' of 'the Conquerour' (388); and the Irish custom of transhumance, or 'Bolloyinge' (1532) – denigrated as one of three 'Scithian abuses' (1664) – results in incessant spatial redefinition, subjecting the surface of the land to continual metamorphoses.

This perspective on Irish space takes as the limits of its visual field the geographical parameters provided by the Irish coastline, it produces the country as a historical, cultural and textual entity without setting it into spatial relationship with the place Irenius and Eudoxus speak from, England. Ireland is viewed in chorographic isolation, not as part of a geographic network, while its landscape, incessantly moralized, is underwritten by a distinct 'ideology of place'[29] that charts the relationship between people and land as inherently transgressive. This symbiotic link, the text suggests, needs to be obliterated, for if the Irish continue to withdraw into a different space, a different *kind* of space, moving them out of reach of the English, proper political and cultural reform is impossible. Irenius' solution is to recommend the 'short, sharp shock' of 'a ruthless war of attrition'[30] to outdo Irish resistance and clear the land for proper recolonization. Once this military proposal has been outlined, highly particularized places and topological constructions yield textually to plans for a massive reorganization of Irish space intended both to implement an effective system of political surveillance and to increase Ireland's financial revenue. It is here that the visual assistance of the map becomes crucial, for Irenius' surveying eye now encompasses in one vista the entire insular topography. This 'view' homogenizes the fragmented Irish landscape by replacing the large-scale representation of cultural aberrance with the small scale of the geographer's canvas, a shift that requires as its enabling condition a systematic process of measuring and quantifying, exemplified by statements such as Irenius' request 'to Consider how much lande theare is in all Vlster' and to estimate the exact 'quantity thereof' (3948–9).

In helping to restructure Irish space in this fashion, the map aspires not merely to describe but to substitute the land on which Irenius' vision is enacted. Much like Tamburlaine's global excess, this transformative cartography generates a new spatial order and annihilates a landscape of custom and use. Turning land into paper no longer restricts Irenius to the narrow, earth-bound perspective of historical chorography but allows him to assume a lofty viewpoint from which he can invent a version of Ireland's future, smooth out the irregularities of Irish topography, and reconfigure its cultural landscape. His *view*, that is, changes from the isolated perspective of the historian, concerned with a specific territorial location, to the scale-model of the political geographer, who shapes the larger structural network of a terrain, embracing both the object of his discourse and the place of his own textual voice. The inherent cartographic logic that governs the *View*'s proposal for political reform thus produces the condition for a spatial fiction which aims to lock Ireland as a former site of cultural difference into its cartographic destiny of geographical proximity to England which necessitates cultural conformity, or 'a vnion of manners and Conformitye of mindes' (4769), as Irenius puts it.

This move – enforced cultural assimilation through spatial homogenization – is the central theme of another contemporary Irish tract, Sir John Davies' legal treatise *A Discovery of the true causes why Ireland was never entirely subdued* (1612).[31] Written almost a decade after the final surrender of Hugh O'Neill the text looks back on a decisive phase in Irish history which saw the influx into Ireland's northern regions of thousands of English and Scottish settlers of Protestant extraction. Davies first examines the reasons why Ireland has not been fully conquered earlier, blaming England explicitly for its insufficient military commitment. More centrally, however, Davies identifies the ill-advised negligence in introducing English law into Ireland as the crucial reason for the incomplete conquest: 'This then I note as a great defect in the Ciuill policy of this kingdom, in that for the space of 350 yeares at least after the Conquest first attempted, the English lawes were not communicated to the Irish, nor the benefit and protection therof allowed vnto them,' adding in truly colonial style, 'though they earnestly desired and sought the same' (116). Since English laws were not enforced, Irish customs prevailed and continued to shape people's lives. Therefore, 'if wee consider the Nature of the Irish Customes, wee shall finde that the people which doth vse them, must *of necessitie* bee Rebelles to all good Gouernment, destroy the commonwealth wherein they liue, and bring Barbarisme and desolation vpon the richest and most fruitfull Land of the world' (165, my italics).

Davies' text subscribes to a spatial logic similar to that which governs Spenser's *View*. For instance, Davies explains that Poynings' laws could not spread throughout the whole island because 'the *Irish Countreyes* [beyond the English pale], which contained two third parts of the *Kingdome*, were not reduced to Shire-Ground, so as in them the Lawes of *England* could not possibly be put in execution' (236, wrongly paginated 220). Irish space lacks subdivision into shires and hence constitutes a category alien to English systematization. Under James, whose reign the fabulous historiography of the *Discovery* is explicitly meant to celebrate,[32] the necessary restructuring of Irish space has now been achieved and 'the whole Realme . . . diuided into *Shires*' (266). In a related example Davies notes how 'the *Mountaines* and *Glynnes* on the South side of *Dublin*, wer lately made a Shire by it self, and called the County of *wicklow*; whereby the Inhabitants which were wont to be Thorns in the side of the *Pale*, are become ciuill and quiet Neighbors thereof' (266). The political argument of the *Discovery*, such passages demonstrate, is organized around one unifying principle: a universally valid legal code with the power to eliminate social, political and cultural difference, making law rather than military force 'the major tool of a practical colonialism in Ireland.'[33] In the *Discovery*, law is a figure of perfect unity, and as the main implement to force Irish realities into English norm and form it turns Ireland into the blank, indifferent surface produced by the measuring operation of the surveyor, ready to be put to its proper use by adventurous settlers. Land in the *Discovery* is a plantation ideal, a projection of civilized English rationality onto inherently anarchic territory, definable in the detached and abstract geometric code of the map.

This vision has a distinctly national dimension. The text is pervaded by a sense of common ground, inhabited by one people 'of English race and Nation' (110 *et passim*). In this unified British territory 'we may con-ceiue an hope, that the next generation [of the Irish], will in tongue & heart, and euery way else, become *English*; so as there will bee no dif-ference or distinction, but the Irish Sea betwixt vs' (272). This vision allows for geographical but not cultural separation and was eventually given cartographic expression by John Speed on his map of *The King-dome of Great Britaine and Ireland* (1611), the opening statement of his *Theatre*, where he presents, with the help of Lythe's old surveys, an image of the British Isles in perfect spatial harmony (Plate 4). Davies welcomed the inclusion of Ireland in Speed's atlas with a dedicatory poem that celebrated the visual access his Irish map provided to 'faire *Hibernia*, that Western Isle', now entirely stripped down to the bare facts of its topography and so thoroughly 'Anatomize[d]' in 'euery *Member*, *Artire*, *Nerue*, and *Veine*, / ... That all may see each parcell without paine.'[34] The *Discovery* is governed precisely by this trope of enhanced visibility achieved through anatomical 'dissection'. As a necessary precondition for Irish cultural conformity space quite literally needs to be cut up into functional 'plots' and redefined within the representa-tional parameters of a political landscape. The incorporation of Ireland into a unified national territory, made possible by the instrumentaliza-tion of law as its enabling cultural matrix, may be a fantasy closer to Davies' than to Spenser's historical moment but informs both texts to different degrees. As in the *View*, the *Discovery*'s vision of an anglicized Ireland is assisted by the advanced representational technology of geometric surveying. Irish alterity, like the social space of rural England, is effaced by writing it off the land, through systematic 'measuring and plotting'.

The conceptual redirections in the attitude to land that generated the paradigmatic shifts in the practice of estate surveying in England are thus replicated, I suggest, on a grander scale in the model of Irish po-litical geography. In both instances the insistence on close social sur-veillance finds an analogue in the application of cartographic scale, and the process of homogenizing space, observable in rural England, is extended across the Irish Sea by the assimilationist rhetoric that informs the textual topographies of Davies' *Discovery* and Spenser's *View*.[35] On the final pages of this chapter I want to briefly turn to a text whose spatializing strategies will be examined in more detail later, Edmund Spenser's monumental poem *The Faerie Queene*. Having started with a scene from a popular contemporary play, my intention is to end by locating the issues discussed so far within the imagery and discursive structure of a self-consciously 'national' literary text conceived and written, away from queen and court, by an exiled English poet in Ireland.[36] The argument, in brief, is to show how the plea for the systematic cultural re-inscription of land observable in the *View* receives wholesale support in the final canto of Book 5 of Spenser's poem but is marked as a doubtful, perhaps unattainable, objective in the first Canto of Mutabilitie – the two passages in *The Faerie Queene* where an allegorized version of contemporary Ireland provides the topographical setting.

When Artegall, the knight of justice and titular hero of Book 5, arrives at the shores of Irena's enthralled kingdom his first view of Ireland is blocked by 'great hostes of men in order martiall' (5.xii.4),[37] sent by the tyrant Grantorto, who gather at the coast to prevent his landing. Artegall's companion, the iron man Talus, quickly dissipates the troops and the knight of justice safely sets foot on Irish soil. But before they reach a nearby city Grantorto's hostile followers reappear in even greater force only to be so severely defeated by Talus that 'they lay scattred ouer all the land, / As thicke as doth the seede after the sowers hand.' (7). Talus, the sower of death, transforms the land into a battlefield covered with the bodies of slain rebels. This image of a wild landscape merging into a thick layer of corpses that literally hide the surface of the map and visually obstruct the cartographic view, is not an isolated occurrence in *The Faerie Queene*.[38] In the context of Artegall's Irish quest, I would suggest, it functions as a reminder that space is not a straightforward, transparent object of description – the liberation of Irena necessarily entails a liberation of the physical space Irena allegorically signifies. As we have seen, landscape is directly the object of military concerns in the *View*: the Irish countryside needs to be laid open, made visible and placed under systematic surveillance, hence Spenser's repeated pleas for it to be 'controlled, shut down, contained, by means of clearings, bridges and forts.'[39] On the battlefield created by Talus in Book 5, this 'cleansing' of Irish landscape takes as its immediate target the social dimension of physical space: it equates the impenetrable bogs, woods and mountains of the *View* with the native inhabitants concealing and 'corrupting' the innocent soil.

After the showdown with Grantorto and the liberation of Irena, Artegall sets out, clearly impressed by the aptness of Irenius' 'plot', to 'reforme that ragged common-weale' (26) and to punish those that 'vsd to rob and steale, / Or did rebell gainst lawfull gouernment' (26). This 'lawfull gouernment', based on English justice finally brought to Ireland by Artegall, needs to be enforced spatially by Talus who recovers for English re-inscription the land transformed into political waste ground by Irish rebels. His mission 'through all that realme' (26), a systematic, quantitative exploration of the land, is not only conceptually akin to the 'perambulation' of the surveyor, it also employs, in the power of Talus' sight 'to reueale / All hidden crimes' (26), the penetrating quality of the surveying eye to cut visually through a landscape disfigured by exposure to savagery. Though interrupted by Artegall's premature recall, a thinly disguised critique of Elizabeth's Irish policy, the initial success of Talus' clinical surveying shows that during his brief presence in Ireland political reform was clearly within reach.

Generally, images of land surveying are woven into the narrative of the poem in a manner unambiguously advocating the necessity for such spatial *re*-vision. In the closing cantos of Book 1, the Redcrosse knight, having just been awarded English name and nationhood in the House of Holiness (1.x.61), ends his quest by confronting a dragon so huge in appearance that, in the manner of Milton's 'sea-beast' Leviathan,[40] his physical greatness almost defies the powers of human perception: 'His blazing eyes, like two bright shining shields, / Did burne with wrath, and sparkled liuing fyre; / As two broad beacons, set in open fields, / Send forth their flames farre off to euery shyre, / And warning giue that enemies conspyre, / With fire and sword the region to inuade' (1.xi.14). Beacons, in Elizabethan times, were a military warning device set on hills along the southern seaboard. In Spenser's image they function as the 'blazing eyes' of the dragon, the substance of whose body extends through the hills into the land itself, here cast in its double status as farmland and administrative pattern, as field and shire. This identification between dragon and land, with its origin in the Ovidian tale of Cadmus' slaying of the dragon, is taken further than the traditional affirmation of the superiority of properly tilled soil over uncultivated land would strictly necessitate, and within the boundaries of the field or the shire the dragon is not merely linked to the land's surface, his huge body almost completely covering the ground, but imagined as its constitutive principle. Immediately after the dragon has been slain by a revitalized Redcrosse, some 'bold' bystanders surround the dead carcass with the intention 'to measure him' and '[t]o proue how many acres he did spread of land' (1.xii.11). Significantly, the activity of measuring, a deliberate intervention in a formerly recalcitrant and uncontrollable landscape, sets in the minute the threat has vanished – a textual move reminiscent of Hooker's Irish chronicle where the disappearance of the rebel ushered in the English land-measurer. Again, as if in explicit analogy to Rathborne's frontispiece, myth gives way to science.

In contrast to this sequence, where the measuring operation announces the necessity of full-scale reform, the first Canto of Mutabilitie subjects Irish topography to a mythical agenda that reverses the order of this episode and recalls the anxieties generated by Irish space recorded in the opening lines of the *View*: it is 'muche to be feared', warns Irenius, that 'the *very Genius* of the soile' or 'some secrete skourge, which shall by her [Ireland] Come vnto Englande' (12–16) are the reasons why nothing prospers in Ireland, no reforms prove successful and civilization remains beyond reach. The first Canto of Mutabilitie expands the transgressive link between land and people into a general account of Ireland's present state of savagery. Interrupting the story of Mutabilitie's trial presided over by veiled Nature, the narrator digresses into a mythical tale involving the goddess Diana whose partiality for its once unspoiled countryside made her select Ireland as a favourite earthly retreat. A conspiracy between Diana's chief nymph Molanna and the foolish wood-god Faunus, who desired to see the goddess naked, tragically results in Diana's departure from earth, 'full of indignation' (7.vi.54) at this rash act of voyeurism. Laying a 'heauy haplesse curse' on the land she has decided to abandon, she decrees that 'Wolues, where she was wont to space, / Should harbour'd be, and all those Woods deface, / And Thieues should rob and spoile that Coast around.' As a result, 'to this day' wolves and thieves abound in Ireland, '[w]hich too-too true that lands in-dwellers since haue found' (55). Acting as a degenerate Irish Actaeon, Faunus – in contrast to the tale's most immediate Ovidian source – causes not his own but Ireland's downfall,[41] and the country's resulting decline into barbarism is made manifest in the state of its physical landscape. The spatial markers in this stanza – 'space', 'harbour', 'Woods', 'Coast', 'Chase', 'lands' – define the land itself as the prime locus of incivility, emphasizing Spenser's predominantly spatial conception of Irish barbarism. Ireland is first of all a wild and inhospitable landscape, a geographical terrain cursed by the gods and infested with social outcasts, removed from the lucidity of the cartographic view. Covered by hordes of wolves and thieves spoiling its erstwhile splendour, Irish landscape must be newly cleared and purified from this corrupting influence to enable its proper reform; as we have seen, it is this course of action which Artegall and Talus decide on in Book 5.

But in the first Mutabilitie Canto, no such effective reform is in sight. Here, the visual power of Talus' reforming crusade, enabling him 'to reueale / All hidden crimes' (5.xii.26), is not only linked to the titular keyword of Spenser's 'plot' and the professional, instrument-supported 'view' of the surveyor, both essential for Ireland's reform, but also, importantly, to its transgressive inversion in the voyeuristic act of Faunus, to whom was unlawfully revealed what should have remained hidden, Diana's naked body. The causality of this sequence, a porno-graphic view resulting in heavenly denunciation, locks Irish landscape in its barbarous state and removes it conceptually from the visual field of the surveyor by casting the critical 'view' which needs to precede Ireland's restoration as doubly the object of fundamental scepticism, marked first as unlawful transgression, then as divine curse. Both these 'views' focus on Irish land as the object of contrasting desires. The first – the voyeurism of the 'foolish *Faune*' (7.vi.46), who might see Diana's 'priuity' (42) once but under penalty of exile – only causes the depar-ture of the goddess which seems as final as the disappearance of the Graces from Colin Clout's Arcadian retreat in the previous Book. In the second instance, Diana, who 'hated sight of liuing eye' (42), casts a last surveying glance over the landscape that betrayed her under guise of the naiad Molanna: '[Diana] quite forsooke / All those faire forrests about *Arlo* hid, / And all that Mountaine which doth over-looke / The richest Champian that may else be rid, / And the faire *Shure*, in which are thousand Salmons bred' (54).

Zooming in from the general to the particular, from the sumptuous panorama of all 'those faire forrests' to the individual detail of the fish in the river, this celebration of the Munster countryside[42] merges into an idyllic pastoral setting, '[t]he richest Champian'. As a view that reg-isters and catalogues the surrounding landscape elements from the sur-veyor's ideal vantage point of Arlo Hill it is reminiscent of the spatial synthesis of the map.[43] Yet this map – unlike Eudoxus' military chart, essentially a chorographic view – vanishes together with Diana; it does not serve to initiate a programme of systematic reform as it did in the *View*, or as it might have done in Book 5. In a move that recalls the egalitarian giant of Book 5, Canto 2, whose intervention in landscape was deemed unlawful since only gods may reshape what has been affected by change,[44] this sweeping cartographic vista, and the survey which might script it on paper, are reclaimed as the exclusive privilege of the gods. The excursion by the 'poet historical' into the origins of Ireland's barbarity thus acknowledges the continuing intractability of a pre-cartographic landscape whose utter decay and geographical dissociation threaten to subvert the rigorously enforced politics of cultural and spatial assimilation. There is another point to be made here, one that concerns not just surveying but the epistemology of space in *The Faerie Queene*. The poem, unlike the *View*, is deeply anti-cartographic; maps are not only absent from the text – Britomart, for instance, needs to find her way '[w]ithouten compasse, or withouten card' (3.ii.7) – they are unimaginable shortcuts that render impossible the collective experience of space and undermine the didactic project of the poem as a whole, the moral education of knight and reader. These spatial anxieties will be considered in detail in Chapter 8; for now I merely wish to register this ambivalence and point to Spenser's trust as well as to his disbelief in the reforming and literally 'ground-breaking' potential inherent in maps and surveys.

To be sure, we are back, if only briefly, in the world of the *View* when the opening stanza of Diana's tale gestures at the ancient spatial harmony of an insular world where Ireland once had pride of place: 'Whylome, when IRELAND florished in fame / Of wealths and goodnesse, far aboue the rest / Of all that beare the *British* Islands name / The Gods then vs'd (for pleasure and for rest) / Oft to resort there-to, when seem'd them best' (7.vi.38). The adjective 'British' here defines the geographical entity of an archipelago but also suggests the political dream of Irenius' 'plot', the recovery of ancient unity in space. In removing this aspiration from the realm of human agency in the ironic Ovidian tale of Ireland's decline into barbarism, while affirming it during Artegall's brief sojourn in Irena's kingdom, both these episodes from *The Faerie Queene* reflect in inverse order the spatial concerns of the *View*: if the political tract aims to replace the larger chorographical with a smaller geographical scale, and thus open the land for cartographic revision, the poetic version of Ireland as a cursed romance setting places the land outside the domain of the map sequentially *after* Talus had set out on his cartographically inspired reforms. The distinction at work here, between the conception of land as a specific locality defined in terms of the people making it 'their' space, and the relational approach of the geographer whose application of geometry synthesizes space beyond its social meaning, can be traced across the textual representation of land in English surveying manuals and in descriptive accounts of Ireland. Individuals in the act of resistance or disobedience – a farmer railing against a surveyor, Irish rebels opposing the crown, Faunus catching a glimpse through the wood – are properties of a pre-cartographic landscape, and ultimately erased from the land organized by the map. In each case the social dimension of space becomes the object of a surveying eye whose 'too seuere scrutations, examinations, impositions, & imputations',[45] while intending to lay open topographical 'truth', explore the full semantic potential of the 'lie of the land'.

Part II
Cartographies

Introduction

In the *Universal History of Infamy*, Jorge Luis Borges included a brief sketch on 'the exactitude of science'. Disguised as an authentic fragment from a seventeenth-century travel account it tells the story of a distant empire where

> the craft of Cartography attained such Perfection that the Map of a Single province covered the space of an entire City, and the Map of the Empire itself an entire Province. In the course of Time, these Extensive maps were found somehow wanting, and so the College of Cartographers evolved a Map of the Empire that was of the same Scale as the Empire and that coincided with it point for point. Less attentive to the Study of Cartography, succeeding Generations came to judge a map of such Magnitude cumbersome, and, not without Irreverence, they abandoned it to the Rigours of sun and Rain. In the western Deserts, tattered Fragments of the Map are still to be found, Sheltering an occasional Beast or beggar; in the whole Nation, no other relic is left of the Discipline of Geography.[1]

This short tale reverses the operative order of cartographic semiosis: blurring the distinction between the abstraction of the image and the substance of the real, the empire's cartographic double transcends its strictly representational purpose and rivals the land as an ideal living space. The adherents of the 'Discipline of Geography', spurred on by a desire for technical perfection, effectively offer to trade in the experience of landscape for its synthetic counterfeit, the scale-map. The sheer scope of the project – a map that covers the land city by city and province by province, that simultaneously multiplies and enwraps the empire – ultimately renders absurd the map's referential function, revealing the hyperbolic gesture as both a measure of cartography's excessive hubris, of its vain bid to rewrite the world as a geometric script, and as a mark of its failed aspirations, of a challenge lost to the superior attraction of a physical world. The fantasy of scalar representation is pushed to an illusory extreme – the perfect equation between imitation and reality – only to have the concept of the map fall into oblivion, its last relics surviving in the margins of its own *topos* to cater for the beasts and beggars at the fringes of the social realm, inhabitants of end-zones and wastelands, locked in metaphorical kinship with the ethnographic marvels of early modern map-makers, cannibals and amazons. The art of cartography may have simply become deeply unfashionable – the disdain later generations show for the efforts of their forebears – but the story hinges on the conceptual confusion between map and land, each the mirror image of the other: the imperial map is cast as a geographical icon whose representational strategies actively create, rather than passively reflect, the actuality of its immediate referent, the space of the land. As 'Mein Herr' explains, in Lewis Carroll's tale which may have been Borges' source, 'we now use the country itself, as its own map, and I assure you it does nearly as well.'[2]

In the first decade of the seventeenth century the fictional farmer in John Norden's *Surveyor's Dialogue* unwittingly raises the same point. '[W]e poore Country-men', he tells the surveyor, 'doe not thinke it good to haue our Lands plotted out, and me thinks in deede it is to very small purpose: for is not the Field it selfe a goodly Map for the Lord to look vpon, better then a painted paper? And what is he the better to see if laid out in colours? He can adde nothing to his land, nor diminish ours: and therefore that labour aboue all may be saued, in mine opinion.'[3] The farmer here argues what he sees as the utter pointlessness of cartographic representation, a useless and futile gesture, which can neither 'add to' nor 'diminish' knowledge of the land. Yet the comment is pervaded by precisely that fear – that the estate map, with the added colour key distinguishing different forms of land use, might change the true nature of the fields it graphically displays and raise unjustified expectations on the part of a profit-oriented lord. Rather than look at the 'painted paper' in his study and speculate on his land's financial value, the lord should honour the field itself with his physical presence and ocular authority: 'the best doung for the feelde', a contemporary adage had it, 'is the maisters foote'.[4] This is a common complaint in an age where increasing economic pressure causes absenteeism among landowners whose role of symbolic fatherhood on the estate yields to rigid commercial management.[5] But the crucial point, for the purpose of my study, is the reliance of the farmer's arguments on a mode of perception that already, before the surveyor even unpacks his instruments, conceives of land in cartographic terms, that effectively recognizes the 'Field it selfe' as nothing but 'a goodly Map'. The farmer – in the manner of Borges' cartographers – translates the immediacy of his surroundings into the parameters of a universal map which challenges, on the scale of one to one, its status as a mere copy of reality. In both accounts the physical world blends in imperceptibly with the cartographic display, and cognitive possession of space, at the levels of empire and estate respectively, is claimed not by first-hand acquaintance with the territory but by the prime shapers of mental images of space, the representational codes of the map.

I offer these passages from Borges and Norden (in their, respectively, ironic and apologetic modes) as textual echoes of a phenomenon which historians of early modern cartography habitually refer to as the emergence of 'map consciousness', the ability to absorb and transmit spatial information encoded in the representational patterns of cartographic projection.[6] The career of the estate map, briefly discussed in Chapter 2, is just one example of the increasingly inseparable link between the imagery of topographical maps and the perception of physical space in early modern Britain. Arguably the most influential instance of this twofold exchange was the power of the scale-map to offer a persuasive emblem for the political space of the nation. Christopher Saxton's atlas of England and Wales, published in 1579, allowed contemporaries for the first time to '[take] effective visual and conceptual possession of the physical kingdom in which they lived',[7] and John Speed's Jacobean atlas of 1611, *The Theatre of the Empire of Great Britain*, expanded this national vision to encompass a unified Britain, referred to in biblical hyperbole as 'the very *Eden of Europe*'.[8] This ideological impact of cartographic representation is a reminder of the central position geographical discourse occupies in the national imagination. On the national map the descriptive paradigms I explored in Part I – cosmography and modern estate surveying – are at once contained in, and superseded by, a cartographically constructed image of sovereignty, simultaneously conceived of as a world by itself and a perfect and well-kept estate. The map of the empire and the map of the individual field, the global and local, act as both imaginary and geographical points of reference for the mapping of national space in early modern Britain.

The ideological frictions generated by this conceptual purchase on national space, made possible by early modern cartography and fostered by 'map consciousness', are my main concern in Part II. Decorative Renaissance maps, of which many early English maps are examples, are often discussed as objects of an almost sensuous attraction whose expansive cartographic scene – taking in 'the whole world at one view'[9] – serves as an index of contemporary spatial desire. But the different maps of Britain that were historically available also illustrate that their elevation into icons of nationhood entailed the complex debate of exactly how to imagine the nation's external shape and internal order. Speed's Edenic image of a British 'garden', for instance, is challenged by the proximity of an Irish 'wilderness'; and as English cartographers set out to define in their maps the space of the nation they were confronted with Ireland's 'constitutional anomaly':[10] neither properly a kingdom nor fully a colony, its political status, and position within the national framework, was fraught with ambiguities. Extant maps – both printed and manuscript – contain a range of suggestions as to how Ireland might relate to the emerging nation, varying from full absorption to deliberate exclusion.

This spatial ambivalence is also demonstrated by the slippages that could occur in contemporary references to national geography. In John of Gaunt's vision of a 'sceptred isle',[11] perhaps the *locus classicus* of such geographical distortions, divine favour and cartographic vision combine to produce the 'island' of England as a 'blessed plot',[12] effacing all traces of Scots and Welsh.[13] In a gesture at once more expansive and even less interested in geographical fact, William Cuningham declared that 'vnder the name of Englande, I comprehend the whole Ilande conteyning also Schotlande, & Irelande'[14] – a memorable passage in a book entirely concerned with technical precision in measuring land, reading the sky and producing accurate maps. Cuningham's casual reference to an all-inclusive English insularity hardly echoes widely shared assumptions about Britain's physical topography but indicates both the imaginary potential and the political relevance of early modern geographical thought. The nexus of the cartographic and the national is ample evidence that maps could – and did – serve many divergent, even contradictory interests. My readings in the chapters that follow will acknowledge this cartographic partiality by starting from the premise that maps never accomplish their representational work in a realm of objective transparency but always refer the viewer back to the cultural and political contexts from which they emerge.

In Part I of this study I discussed the discursive erasure of social space inherent in the moment of *measuring* with reference to three distinct contexts – the intellectual appropriation of global space in cosmographical thought, the arrival of the modern mathematical surveyor in the social world of rural England, and the spatial logic of political reform plans in Ireland. Part II turns to the instant where such spatial reconstructions are directly *visualized* as cartographic images. The first chapter looks at contemporary perceptions of cartography. Focusing on the ambiguous function of maps as spatial signifiers in early modern times, I will discuss the cartographic image as a visual site on which territorial and mental 'views' overlap and mutually enable each other. The second chapter looks at the maps of Laurence Nowell, Christopher Saxton and John Speed to examine the map of England (and/or Britain) as both an aesthetic and a political experiment on the idea of the nation. These were not the only 'national' maps available.[15] But in contrast to the maps included, for instance, in the atlases of Ortelius and Mercator, the maps of Nowell, Saxton and Speed were all produced within a recognizably English frame of reference; their cartographic activity bears directly on the internal perception of national space in early modern England. The final chapter looks at examples of Irish maps that had various functions (ornamental, strategic, legal), and which all resonate with wider political and cultural tensions. In gradually redefining the savage, intractable Irish wasteland as a territorial extension of the national sphere, these maps negotiate the political accommodation of Irish difference into a 'British' framework. But while maps are suggestive records of such tensions they do not exist in isolation, and a similar visual code, I argue, generates the fluid spatial imagery prevalent in dramatic representations of the 'saluage Iland'.[16] It is the parallel function of map and stage in providing suitable images of Ireland, the discursive connection between the theatrical and the cartographic, I intend to explore by linking Irish maps to the plays of Shakespeare, a prime textual site that addresses, if only indirectly, the Anglo-Irish confrontation of the sixteenth century.

4

The Whole World at One View

Maps, it has often been noted, were the objects of considerable enthusiasm in early modern England. 'I Daylie see many that delight to looke on Mappes',[1] wrote Thomas Blundeville in 1589, echoing John Dee's famous comment, then almost two decades old, that 'some, for one purpose: and some, for an other, liketh, loueth, getteth, and vseth, Mappes, Chartes, & Geographicall Globes.' As Dee explained, the purposes to which these geographical artefacts could be put were various: as items of interior decoration they served 'to beautifie . . . Halls, Parlers, Chambers, Galeries, Studies, or Libraries', as pedagogic study aids they could illustrate important historical events, and the practical value of their topographical information would enable journeys 'into farre landes: or to vnderstand of other mens trauailes'.[2] Such quotations reflect the immense upsurge of interest in geographical maps that prominently articulated itself in the sixteenth century.[3] The wide (and fairly sudden) acceptance of cartographic images is usually read as indication of a growing 'map consciousness', referred to above, that began to conceive of the world as written in the geometrical language of maps and charts. As tools of government, as practical guides directing a fast growing fleet of merchants around the globe, as prestigious objects displaying a landowner's wealth and as powerful visual conceptualizations of the 'imagined community of the nation',[4] maps of various types and genres[5] were beginning to enjoy such a high degree of circulation in sixteenth-century England that modern historians are regularly prompted to diagnose a 'cartographic revolution'.[6]

As early as 1531 Sir Thomas Elyot praised maps for their didactic advantage over the written word and their ability to set historical events into spatial contexts. 'Also to prepare the childe to vnderstandynge of histories,' he writes in *The Book Named the Governor*, whiche beinge replenished with the names of countrayes and townes vnknowen to the reder, do make the historie tedious, or els the lasse pleasant, so if they bee in any wyse knowen, it increaseth an inexplicable delectation. It shall be therfore and also for refreshing the witte, a conuenient lesson to beholde the olde tables of *Ptholomee*, where in all the worlde is paynted.[7]

Elyot significantly explained the attraction of the universal map of the world, a painted image of the earthly globe reconstructed from the rules laid down in Ptolemy's *Geography*, not in terms of accuracy or precise geographical content but as resulting directly from its ideal representation of earthly *copia* and its ability to enable imaginary journeys through space and time:

For what pleasure is it, in one houre to beholde those realmes, cities, seas, ryuers, and mountaynes, that vneth in an olde mannes life can nat be iournaide and pursued: what incredible delite is taken in beholding the diuersities of people, beastis, foules, fisshes, trees, frutes, and herbes? To knowe the sondry maners & conditions of people, and the varietie of their natures, and that in a warme studie or perler, without perill of the see, or daunger of longe and paynfull iournayes? I can nat tell, what more pleasure shulde happen to a gentil witte, than to beholde in his owne house euery thynge that with in all the worlde is contained.[8]

In sixteenth-century Europe, the notion that maps could transport the universal variety of the world into the domestic orbit of the armchair geographer was a fantasy shared by many, and comments of this nature are ubiquitous in contemporary accounts of map reading. The kinship of the man of letters, contemplating the entire world in 'his owne house', with the lord of the manor, gazing at his lands while 'sitting in his chayre at home',[9] reminds us of the way estate, national and, for Elyot, world maps cultivated what is essentially a rhetoric of property. All radiated the implicit conviction, however imaginary, that ownership of the map laid a claim to the ownership of the land, a proprietary relation understood as much in mental as in material terms.

Although the popularity of the Ptolemaic world map – Elyot's point of reference – was soon eclipsed by the map collections of Ortelius and Mercator, fashionable products of the 'new geography', this technical revolution hardly affected the reasons on which the appreciation of cartography was based. Robert Burton, writing almost a century after Elyot, praised modern world and city atlases in similar terms:

> To some kind of men it is an extraordinary delight to study, to looke vpon a Geographicall mappe, and to behold, as it were, all the remote Provinces, Townes, Cities of the world, and never to goe forth of the limits of his study, to measure by a Scale and Compasse, their extent, distance, examine their site, &c. What greater pleasure can there be then to view those elaborate Maps of *Ortelius, Mercator, Hondius*, &c. To peruse those bookes of Citties, put out by *Braunus*, and *Hogenbergius*.[10]

If Elyot's ethnographic interest in 'the diuersities of people', their 'sondry maners & conditions', clearly gives way – in analogy to the changes examined in Part I – to Burton's concentration on quantifiable space, on the 'extent' and 'distance' of places to be measured from within the confines of the gentleman's study, both statements still share the exuberant praise for a mode of pictorial representation which allows the hazards of travel to be replaced by the comfort of solitary contemplation. Separated by nearly a century, both also reflect the contemporary understanding of the interrelation between space and time which rationalized the knowledge of geography as a necessary adjunct to the lessons of history. Just as historical events were grounded in space, so their geographical setting served as a backdrop to the timescale of history. These views thus attribute to maps a function that merges illustration with interpretation, indicating that they were more than mere ornamental wall-hangings or purely the leisurely pursuits of a social and intellectual elite (although they frequently figured as such). Rather, they are instruments of new taxonomies, highly flexible means of classifying knowledge about the world, which processed topographical, ethnographic and political information and offered it up to human eyesight in the concentrated form of geographical images.

The range of the contemporary debate on cartographic matters and the currency of the map as a metaphorical paradigm indicate that geographical maps, particularly those found on walls and in the pages of an atlas, were agents of new cultural and ideological forces in Elizabethan and Jacobean times. Throughout history, of course, maps have always been evidence of the desire to transfer the perception of the landscape one was inhabiting into some kind of representational image but it was in the sixteenth century that mapping entered everyday life in largely unprecedented ways.[11] On their respective content

levels – estate, county, nation, world – maps organized space by combining the relics of symbolic spatial hierarchies inherited from a
medieval tradition, iconic investments in the inner quality of space,
with an empirical gesture that measured and charted its quantity, its
precise width and breadth. Despite their self-proclaimed ease in evoking,
indeed imitating, a tangible world 'out there', maps are never 'natural'
images of physical space but carefully edited, multi-layered topographical views generating subtleties of meaning that far exceed any mere
illustrative value. J. B. Harley, whose writings have most persistently
argued this point, claims that Tudor maps articulate

> symbolic values as part of a visual language by which specific inter-
> ests, doctrines and even world views were communicated. Maps were
> one of a number of instruments of control by landlords and gov-
> ernments; they were spatial emblems of power in society; they were
> artefacts in the creation of myth; and they influenced perceptions of
> place and space at a variety of geographical scales.[12]

The high emotional charge of Tudor and Stuart maps of Britain, and the
national agenda they openly promoted, was thus achieved by establishing cartographic discourse as one element in a whole network of
economic and social forces which redefined the meaning of land and
reshaped the actual and symbolic landscape of Great Britain and Ireland.
The finger running along the map, imitating the body moving through
the landscape, fostered wider imaginative circulations round the nation,
turning national geography into an educational programme. John
Norden – admittedly not an impartial observer – advised King James
that 'it well befitteth a Prince to be trulie acquaynted with his owne
territories',[13] indicating with this remark the increasing importance
attached in early modern Britain to familiarity with British geography,
with the genealogies of leading families, or with Britain's place-names
and the exact location of towns, hills and forests – the very agenda of
the chorographical enterprise.[14]

The cartographic gesture of spatial appropriation thus expands the
perceptual field of topography and reaches far beyond the factuality of
landscape to articulate geographical data in terms of a cultural and political script. This accounts for the synthesizing effect of maps, their impo-
sition of visual coherence on the physical shape of the earth, but also
for their emphasis on the disruptive elements, the various forms of
'otherness', existing within that representational framework. Maps,
contemporaries were well aware, dealt in facts of cultural difference as

well as in myths of spatial unity. The Protestant zealot William Vaughan, for instance, resorted to map imagery when he described a heavenly voice advising him to get 'an exquisite map of all this *Iland*, and view whether there be not ten Tauernes for one Church, ten diuels for one Saint, ten tospots for one temperate.'[15] The metaphorical map of Britain here shows a society on the brink of corruption, it highlights domestic dangers and the absence of a universally accepted social and religious consensus. Spatial relationships across the globe, as they were brought to consciousness by the cartographic display, could be similarly fraught with political and religious anxieties. Thus John Dee, in the continuation of the passage quoted above, notes how world maps show side by side that 'litle morsell of ground, where Christendome (by profession) is certainly knowen' and the 'large dominion of the Turke: the wide Empire of the Moschouite.'[16] Obliquely reminiscent of the moral economy of an ancient cosmographic tradition, this comment on cartographic proportion adds a cultural and religious dimension to the map's mercantile and navigational function: the reference to the 'Empire of the Moschouite' reflects the economic interests of British merchants but also records the 'large' and 'wide' spaces of ethnic and religious diversity which the small scale of the new world map had now brought into much sharper focus.

If some contemporaries read maps as displays of cultural difference, some thought they invited serious misreadings of political space. A frequently quoted complaint by the Welshman George Owen, chorographer of Pembrokeshire, was directed against the English government's apparent failure to grasp the concept of cartographic scale. For the heavy tax burden which Owen believed to have unjustly fallen on his native county he blamed its cartographic representation in Saxton's atlas where Pembrokeshire was the only Welsh county 'haveinge the rome and place of a whole sheete of paper allowed to it selfe'.[17] On all the remaining sheets of the Welsh section up to four counties competed for the viewer's attention, thus giving the impression of covering together the same extent of space Pembrokeshire occupied on its own. The effect, according to Owen, was that government officials, owing to their apparent ignorance of the concept of scale, thought Pembrokeshire larger in size than it actually was and, consequently, taxed it far more heavily than the remaining Welsh counties.[18] In the context of a 1571 debate on county representation, a few years before Saxton's maps started appearing in print, a parliamentarian alluded to the apparent inefficiency of maps to convey a sense of the true state of the country:

How may her Majesty, or how may this Court know the estate of her Frontiers, or who shall make Report of the Ports, or how every Quarter, Shire or Country is in state? We who never have seen *Berwick* or St *Michael's* Mount, can but blindly guess of them, albeit we look on the Maps, that came from thence, or see Letters of Instruction sent; some one whom Observation, Experience, and due Consideration of that Country hath taught, can more perfectly open what shall in question thereof grow, and more effectually reason thereupon, than the skilfullest otherwise whatsoever.[19]

This view privileges local experience over the abstract information contained in maps and suggests the necessity of grounding spatial knowledge in the direct acquaintance with regional topography. Evidently, not everybody was convinced, as is often claimed, that maps could effortlessly 'telescope the power of human sight, both temporally and spatially'.[20] And if some contemporaries recognized that maps may condense geographical information to the extent of distorting it, their perception of cartographic practice may have been as ambiguous as the map image itself.

Nevertheless, the visual power of maps was still most commonly singled out for praise in contemporary accounts. Arthur Hopton revelled in the fantasy of a perfectly geometricized world, subject to the penetrating clarity of the map:

> Heer's no vaine shew: illusions haue no place,
> No spirit confind, no hatefull painted face,
> No eye-deceiuing glasse, no Cristall braue,
> Which from the frozen seas we often haue.
> But in a faire and most perspicuous light,
> The earthy Globe lies subiect to thy sight.[21]

The direct exchange between map and eye, undisturbed by human displays of vanity, was rationalized as an effective didactic short-cut. '[T]he Mappe being layed before our eyes,' wrote Ortelius, 'we may behold things done, or places where they were done, as if they were at this time present and in doing'.[22] Maps, for Ortelius, achieve the perfect illusion of an immediate, almost mysterious presence, and they 'work' in this way, Denis Wood argues, precisely because they 'give us *reality*, a reality that exceeds our vision, our reach, the span of our days, a reality we achieve no other way.'[23] They manage, that is, to pass off for evident truth what is hard won, culturally acquired knowledge about the world we inhabit, a reality unverifiable by the naked eye: by making us see what eludes our visual perception, by dragging into open view literally invisible spaces, maps promise with each line to transcend the limited powers of human eyesight. The sense of privileged visibility intrinsic to cartographic representation elicited enthusiastic responses from early modern commentators. '[S]tudy well these moderne Maps', Blundeville recommended, 'and with your eie you shall beholde, not onely the whole world at one view, but also euery particular place contained therein.'[24] The generic labels of contemporary atlases, such as the 'glass', the 'mirror' or the 'speculum', equally foregrounded ocular effects; and when Ortelius, in the preface of his bestselling world atlas, called geography 'the *eye* of History'[25] he referred directly to the visual power of maps and their ability to expand the perceptive faculties otherwise limited by the restraints of space and time.

Yet if only to exploit a well known metaphor, it should be noted that some eyes are as blind as others are observant, and contemporaries also recognized that the abstraction of geometric scale may quietly conceal rather than openly disclose geographical information. Trust in the usefulness of maps as visual tools was never as unconditional as Blundeville's eulogy makes it appear. Samuel Daniel, for instance, took the opposite stand when he judged maps to be indistinct and literally superficial images of space: 'We must not looke vpon the immense course of times past', he explained in his *Defence of Ryme* (1603), 'as men ouer-looke spacious and wide countries, from off high Mountaines, and are neuer the neere to iudge of the true Nature of the soyle, or the particular syte and face of those territories they see. Nor must we thinke, viewing the superficiall figure of a region in a Mappe that wee knowe straight the fashion and place as it is.'[26] For Daniel, the poverty of cartographic information is a consequence of the distance between the map and its object of representation. Lacking the immediacy spatial knowledge requires, maps are insufficiently detailed and wrongly focused records of space, vague approximations at best, which concentrate exclusively on the outward 'figure', not the inner quality, of landscape. They are, in fact, instruments of a worrying delusion which may effectively prevent visual access to vital information by keeping the viewer at a distance from the 'reality' of the region depicted, from the 'fashion and place *as it is.*'

The distinction set up here, between maps as either completely revelatory or conceptually blurred images of landscape, should not come as a surprise. It is hard to dispute that maps indeed manage to style themselves as instruments of unlimited ocular inspection, but only after they have first simplified their mode of pictorial composition and selected only certain bits of spatial information as worthy of representation; it is this dual process of internal revision which allows them to convincingly imitate what they casually refer to as the 'real'. On the content level, in the choice of what landscape elements to include in the visual display, conscious editorial arrangements of the cartographic picture plane are most clearly noticeable, and these 'superficial' reductions in spatial complexity are repeated on the level of representational technique. Worsop, the mathematically inclined land surveyor we have already met, explained that cartographers simply cannot be expected to convey an 'accurate' impression of landscape since they are exclusively concerned with 'the vpper face of any thing', 'desir[ing] only to know the content of the outward plaine . . . not regarding thicknes, weight, grossenes or depth: but only the mesure of ye vpper parts as in groundes: which consist onely of length, and bredth, whether they be flats, or leuels, hils, or valleis.'[27] Nevertheless, such graphic minimalism managed to generate the telescopic fantasy of a world fully accessible to the human eye. Evidently, to call a map a selective vision of space endowed with the unique ability to feign a comprehensive image of 'reality' is not to accuse maps of a secret manipulative power but merely to comment on a constitutive aspect of cartographic projection.

Telescopic sight is hardly the bedrock of modern cartography; it is the product of an ideological discourse which relied for its political usefulness on the idea, rather than the fact, of unlimited ocular inspection. Maps undermine, as well as affirm, a straightforward visual code of unrestricted visibility, and the topographical surface they purport to describe is subject to an interplay of visibility and shadow that hides and obscures while claiming to reveal and lay open. Luring the viewer into a seeming familiarity with the landscape on open display, the cartographic picture plane engages in moments of blindness and lucidity that either efface or bring into focus the conceptual and physical distance between the map and the territory it sets out to represent. Maps, that is, construct a conspicuous 'absence' as much as – for Ortelius – a mysterious 'presence'. The enthusiastic response to the visual appropriation of physical space was principally a tribute to the effective coalescence of a powerful visual language with the respective cultural and political formulae endemic in the representational codes of the map. In presenting space as politically and economically manageable, in promising imaginary control through ocular access, cartographic investments prefigured material exploits and the act of beholding 'the whole world at one view' anticipated the imperial gaze of spatial desire.

Maps not only afforded but were themselves 'views', and the centrality of this word in cartographic discourse is perhaps best illustrated through an analysis of its conceptual usage in the context of land surveying. Like other commonly used expressions – perambulation, description, plot, and so on – 'viewing' is a basic surveying term, denoting both an act of visual perception and a technical operation. Before getting out their instruments, Rathborne advises, surveyors should first 'ride or walke abroad, and . . . take *a respectiue view* of the situation and extent of the Mannor', in order to find the best spot to 'perform [the] Instrumentall mensuration.'[28] Bad surveyors, frequently the object of Rathborne's derision, are significantly described as blind and therefore unable to take a proper 'view': 'But what should I say more of them . . . [I will] leaue the blind, with tumbling the blind into the myre.'[29] 'Viewing' has always been part of surveying terminology but did not always refer to the same practice. When Valentine Leigh explained, in 1577, '[h]ow a Surueiour should take a perfecte View or Suruey of a Mannour or such Landes Tenementes or Hereditamentes',[30] local topography was only of marginal interest: 'viewing' referred to the examination of legal documents pertaining to details of tenure. But for Norden, half a century later, 'a due, true, and exact view'[31] already presupposed the physical act of looking through the sights of an instrument. The self-empowerment implicit in this process of 'viewing' is richly embellished by Agas who makes his own eyes the sheer omnipotent centre of the measuring operation: 'And I thinke it a matter both pleasant and profitable, that but once seated in the middle of a quadrangle sumptuously builded about, I may take (by pitching & fastning my instrument onely in that place) all manner of measures there *subiect to myne eye whatsoeuer'*.[32]

Three aspects of this technical act of 'viewing', performed by the 'measurers eye',[33] strike me as particularly relevant. First, it not so much intensified as added a whole new dimension to the power of human sight. Considering the deliberate emphasis on visual aspects in the discourse of land surveying it is surely interesting to learn that the telescope is never so much as mentioned in any of the surveying manuals written up to the mid-seventeenth century. But then the telescope is of a different order conceptually from the principal surveying instrument that defined the profession, the theodolite. The former is an optical instrument, designed to increase the power of the human eye, whereas the latter allows the 'discovery' of hidden, and hence invisible, truths, accessible only 'intellectually', through the application of geometrical knowledge. The theodolite, other than the telescope, was a different kind of eye, which surpassed the faculty of human sight. Second, the surveyor's 'view' could take on various discursive modes that would yield different sorts of information. Norden explains, for example, that 'the want of due plots and descriptions of land in this forme [of the survey], hath bin the occasion of infinite concealements, and losses of many mens land, and many intrusions and incrochments haue bin made'.[34] Identifying concealed space or encroachments has legal and fiscal, rather than topographical, relevance. Third, land was seen not through a perspectival arrangement but through a descriptive grid. The structural pattern of the surveyor's 'view' is the geometric chart; it is not the projection into space of a subjective point of view. Though the surveyor is physically positioned somewhere in the middle of the landscape he does not allow his own eyes to determine the vanishing point of the map.

This scientific streamlining of cartographic 'views' was part and parcel of the ideological power invested in maps, it defined cartographers as specialists with the sheer vatic ability to see the 'truth' about land. The semantic spectrum of the word 'view', as used in surveying, thus spans both formal and visual aspects, and a similar typology undergirds the etymologically related word 'survey'. In both words, a third strand of meaning is implicit, the capacity to denote 'a mental image.' The figurative meaning of 'survey', for instance, is listed in the OED (sense 4) as 'a comprehensive mental view, or (usually) literary examination, discussion, or description, *of* something.' This is a familiar and still common usage of the term, with an obvious structural relation to the activity of the land surveyor. It flows into Norden's own rationalization of his craft but even more crucially applies to cartography in general: maps record both the cognitive perception and the abstract conception of space; they fuse external topographies with a society's dominant political narratives in order to affirm, in the mind of the map-user, an imagined congruence between natural, cultural and social landscapes.[35]

Thus, if maps are 'views', they are images of social relations as much as topographical records. To visualize the lived space of the land in terms of the geometric code of the map was also to articulate, from the perspective of the map-maker or user, the social set-up of the community that inhabited that space. Before turning to maps themselves in the next two chapters, I want to briefly demonstrate the referential complexity of the cartographic sign with a look at what may well be the best known appearance of a national map in English Renaissance literature, the cartographic prop Shakespeare brings on stage in the opening scene of *King*

Lear.[36] When the ageing king calls out 'Give me the map there. Know that we have divided / In three our kingdom' (I.i.35–6), the idiom of political partition vies with the language of pastoral idealism that follows – its sumptuous description of a space filled with 'shadowy forests and with champaigns riched, / With plenteous rivers and wide-skirted meads' (62–3). We thus learn from the start that the cartographic image of unity is only the visual prelude to a process of political disintegration, the 'unfurling of a programme of brutal partition.'[37] This profoundly negative symbolic function of a map, which prefigures the fragmentation of the national realm acted out in the play as a whole, is a reminder that in early modern English drama, 'the stage's version of the national map is generically grim, functional and minimal', the presence of a map on stage – in marked contrast to its patriotic effect in the public world beyond the theatre – is more often than not 'a signal of national decay rather than the celebration of national mystique'.[38]

In the play, the verbal evocation of the map presupposes Lear's visual contact with an actual map used as a stage prop, and the scene revolves around the pointing and gazing at the image of the kingdom.[39] Lear's theatrical 'viewing' of the map offers in quick succession three different versions of the space encoded on the national map. Beginning with the deictic reference to 'all these bounds even from this line, to this' (61) – the land offered to Goneril – Lear proceeds to confer an 'ample third of our fair kingdom' (78) upon Regan. In refusing to award the remaining section to Cordelia, he increases the shares of Cornwall and Albany: 'With my two daughters' dowers digest the third' (126). The word 'dower', used here (in the Folio text) for the third time since Lear has entered the stage,[40] turns the land into a paternal gift presented in the context of a dynastic marriage. In only a few lines, then, we move from a moment where the map image is the material precondition to speak about Britain in any meaningful sense at all ('from this line, to this'), to a conceptualization of space as the result of a mathematical operation (a 'third' of the kingdom), and, finally, to a situation where '[t]he map, and the land it obliquely represents, are caught up in a dynastic transmission of territory'[41] (the land as 'dower'). What I wish to emphasize here is the subtle, protean shift of Lear's terms of reference when gesturing at the space displayed on the national map: the land is either mistaken for the image itself, reduced to a set of geometric coordinates, or instrumentalized in a test of filial obedience. This functional multiplicity of the cartographic sign should warn us that maps are not, as Terence Hawkes has claimed with reference to *Lear*, simply pragmatic charts which 'operate to reduce an imprecisely defined but nonetheless emotionally charged, numinous and multi-level sense of "nationhood" to the merely literal, physical, one-dimensional stand-ing of a piece of paper.'[42] It will be my claim in the following chapter that this argument fails to acknowledge cartography's complex partici-pation in the struggle over the meaning of the nation as an emerging model of collective political identity.

5
Mapping the Nation

The cartographic description of Britain, product of the combined effort of scholar and field worker, should not be discussed merely as a reflection of the growing accuracy of geographical knowledge or the increasing efficiency of modern surveying techniques. Cartography is not evidence of 'a mechanical correspondence between map and landscape'[1] and the early modern map of Britain is centrally implicated in the construction of a spatial paradigm fit to accommodate a national perspective. From the beginning it was also a matter of immediate political relevance. One of the earliest general maps of Britain that resulted directly from the state's increasing demand for maps as administrative tools was Laurence Nowell's *General Description of England and Ireland* (1564/5, Plate 5), 'arguably the first map that cannot be regarded as essentially a derivation from the mid-fourteenth century "Gough" map of the British Isles'.[2] Christopher Saxton's shadow lies heavily on Nowell's efforts whose map has received much less attention than it deserves. Testifying to its political importance, the small pocket map survives in the single manuscript copy formerly owned by William Cecil, Elizabeth's geographically minded principal secretary, whose extensive annotations, spread profusely over the verso, trace various itineraries to the northern English counties and beyond to Scotland.[3] This map, unusually detailed for its time and size, was almost certainly a preliminary specimen intended to promote a larger project that would have included, if executed, a whole series of individual maps covering the entire area of what Nowell described, in an extant letter addressed to Cecil, as 'regionem nostram' – *our region.*[4] For Christopher Saxton, who began work on a regional atlas a decade later, the national survey needed to stretch no further than England and Wales. The precise meaning of Nowell's possessive phrase, 'our region', may ultimately be uncertain but judging from the *General Description* it hardly coincides with the limited scope of Saxton's *Anglia* (Plate 6).

The map covers the Scottish lowlands, and the adjoining coastline of Flanders and France but is principally a representation of the entire territory of the Tudor state, including Ireland; it thus articulates what Steven Ellis has recently called, with reference to the earlier part of the century, the 'collective view of the dominions which [the Tudor monarchs] ruled.'[5] This is in itself an important political statement since large sections of the area covered by the map presented in fact, even if not in official proclamation, considerable difficulties to the exercise of royal control. Cartographically, this political claim is sustained by the prominent display of the crowned Tudor Arms in the top left-hand corner, matched on the right by a decorated title cartouche. The remaining ornamental additions include seven ships (five of them encircling Ireland), one large fish and – most prominently – the portraits of two male figures in the bottom corners, one attacked by a baying hound. To most students of this map, the portraits suggest the existence of a second representational plane, an arena removed from the cartographic display and occupied by two human agents contemplating the theatrical scene of a land fully revealed to the observing eye. This, at least, is implied by the biographical explanation first put forward by Robin Flower in 1935 (and accepted by most recent scholarship on Nowell) that these are images of Nowell and Cecil, with the cartographer on the left in a state of despair and an impatient Cecil on the right censuring his work.[6] This possible, though debatable, attribution (which I shall look at in some detail in the next chapter) is usually supported with the reference to an extant letter Nowell sent to Cecil in 1563, containing his initial offer to prepare a series of maps covering the whole of Britain: 'I truly hope to effect a description of our region not only as a whole but of all its parts and single provinces.'[7]

This passage confirms the considerable relevance of Nowell's manuscript map in the history of the early modern mapping of Britain,[8] an importance not due to its distribution – Cecil, later Lord Burghley, was the only one reputed to have 'carried [it] always about him'[9] – but to the conceptual approach outlined in Nowell's letter: it is usually cited as the first mention of the ambitious plan (later executed by Christopher Saxton) to systematically survey and map all the individual counties of England and Wales with the aim of producing a coherent national atlas.[10] Though widely accepted, this reading runs the danger of projecting backward in time a scheme that was only beginning to be put into practice a decade later. Nowell's reference to 'our region' (*regionem nostram*), it is important to see, leaves unclear the precise extent of this region – does it refer to the British Isles, Great Britain, England and Wales, or the extent of the English monarch's territorial possessions? Nor are the nature of the subdivisions – 'partes' and 'singulas provincias' – any easier to determine. Some cartographic historians, with a view to establishing a linear pedigree for Saxton's atlas, assume that these terms – and especially 'provincias'[11] – refer to counties, but a classificatory system other than this political pattern seems equally possible. Nowell, the Anglo-Saxon scholar, may have been thinking in terms of the Saxon heptarchy, or simply – as cartographers may be expected to do – in terms of the natural geography. Or, perhaps, in terms of a cultural landscape. His map marks county names in red ink but does not delineate them. It does, however, include Ireland and parts of Scotland, both set against a different background colour (yellow) from the patches of light green covering England and Wales. It is surely not irrelevant that Ireland, a prominent section of the Tudor state, is entirely missed out on the maps later produced by Saxton. If Nowell's manuscript map is any evidence for the larger project he had in mind, the possibility must be considered that the reference to 'partes' and 'singulas provincias' does not gesture at a topographical typology of English and Welsh counties but at a cartographic project that was intended to replicate on a series of maps the complex cultural patchwork inscribed on the natural geography of the British Isles.

The possibility of this alternative conception of a national atlas, based not on the county but on cultural and ethnic regionality, should be kept in mind when considering the work of Christopher Saxton who still remains the best known Elizabethan map-maker on account of the publication, in 1579, of the first complete county-by-county atlas of England and Wales.[12] A decade after Nowell presented Cecil with the *General Description* Saxton set out, issued with a pass from the Privy Council, to apply the estate surveying techniques discussed in Chapter 2 to the national territory.[13] The surveyor Saxton who, for a period of five years, went hopping from hill to hill all across England and Wales can thus claim to be the only person in sixteenth-century Britain whose real-life visual experience approximated the comprehensive utopian view of the atlas. The maps resulting from this national survey were first sent to Burghley for inspection, then published as single sheets (over the period 1574 to 1578), and finally collected in one comprehensive compilation. The publication of this collection was a notable event in the history of European cartography and a major commercial success. The book remained untitled and contained 38 different sheets: a group of three consisting of a frontispiece depicting the queen, an index listing the individual maps, and the royal coat of arms together with various tables breaking down the topographical data into separate lists. The introductory matter is followed by the general map showing England and Wales, entitled *Anglia* (Plate 6), and finally by 34 regional maps of different scale depicting 52 individual counties.

All the maps were of unusually high technical quality and it is tempting to speculate about the sheer force of their original visual impact, particularly since the magnitude of the collection and the geographical detail it provides are unique for the time. There exists no real precursor to the collection, no conceptual model that Saxton could have followed. Perhaps he simply adopted the layout of the world atlases which had been coming from the Dutch presses since 1570, as has frequently been claimed,[14] but the reduction in scale from world to nation is hardly as smooth a transition as this suggestion implies. Maps of individual counties, their space historically known and culturally defined as English, are not straightforward conceptual equivalents of Mercator's or Ortelius' maps which charted not the political landscape of an individual nation but the exotic richness and ample *copia* of the entire human sphere. Both projects intersect in the opening map *Anglia* but the absorption of England into a cosmographical survey of the entire world is of a different geographical order from a collection of maps that makes 'England' the umbrella term for a series of detailed regional representations. Saxton's maps were highly popular and his fame and pioneer status quickly established. By reinforcing a sense of regional identity they initiated a transformation in the contemporary construction of the signifier England, becoming for 200 years the models on which nearly all subsequent maps were based.[15] References to the maps abound in contemporary literature, and derivatives were quickly disseminated: they were printed on playing cards;[16] they were used as wall decoration; they illustrated antiquarian and poetic works such as Camden's *Britannia* (from 1607 onwards) or Drayton's *Poly-Olbion* (1612/22); they even appeared on paintings, most famously on the 1596 Ditchley Portrait of Queen Elizabeth. As a regional atlas, the collection remained unrivalled until John Speed published his *Theatre* in 1611, a differently focused national atlas which drew heavily on Saxton's work.

These maps confronted contemporaries with a version of national space that varied considerably from Nowell's. In marked contrast to the *General Description*, *Anglia* conceives of the area covered by the atlas, effectively England and Wales, as an entity unto itself. Relegated to the margins of the image are fragments of Scotland, Ireland and France. If

Nowell's map resonates with an uneasy network of regional difference – the prominent inclusion of Ireland, the itineraries to Scotland noted on the verso, the diversity of its linguistic code – Saxton's *Anglia* shares in a different cultural agenda. Restricting ocular access to England and Wales the map subscribes to a narrower, more selective vision of the national territory but achieves in turn a degree of graphic consistency largely absent from the *General Description*: *Anglia*'s homogeneous landscape answers not to the cultural difference of its population or the geographical specificity of the terrain but to the administrative unit of the county or shire. This horizontal subdivision of the nation, on which the entire atlas is based, decentralizes the cartographic image and preempts the reinsertion of spatial hierarchies. The space of *Anglia* is governed by a 'tectonic code'[17] that organizes, on a scalar model, the horizontal relationship between counties rather than their topological value in external taxonomies. A visually coherent spatialization of the 'imagined political community of the nation',[18] *Anglia* offers the central frame of reference to which the individual maps of the shires metonymically relate.

Saxton enjoyed full governmental backing during his survey and the whole project should at least partly be seen as an intentional propaganda effort on the part of the Tudor monarchy. Elizabeth's image graces the frontispiece of the collection and the patronage system that led to its production can be traced in the hierarchical line that leads, on nearly every page of the atlas, from the cartographer's compass through the heraldic motto of his patron Thomas Seckford to the omnipresent arms of the Tudors. Eventually, as Helgerson has noted, the priority given to the land itself on maps of England contributed to a conception of nationhood that would dispute the ancient dynastic identification between the land and the body of the monarch.[19] The cartographic image of England, from the publication of Saxton's atlas onwards, unintentionally fostered the radical opinion that loyalty to the land and loyalty to the king could be thought of as separate issues. But on *Anglia* the land is still unambiguously marked as the space of the monarch. The sizeable fleet sailing the English and Welsh coastline, the elaborate title cartouche, the safe exclusion of a non-'English' world – all these features serve to construct the political space of the kingdom as a symbolic sphere of national splendour enabled by a powerful monarchy. It is not any pre-existing geographical reality but the 'presentational code' or ordering device of the map which produces, by arranging the cartographic features into the 'architecture of the picture plane',[20] the pictorially coherent and textually articulate surface of Tudor England.

Saxton's map is truly national in scope – a systematic collection of toponyms which mythically unites as potential equals all individuals living in that space. More than simply an illustration of English and Welsh geography it is a description of the land's inherent 'Englishness', the idealized emblem of a 'blessed plot', cut off from the continent, discovered by its chorographers and painted by map-makers.

How is this national space iconographically constructed? From the first map in the collection – *Anglia* – onwards, a dynamic quality is apparent. Rivers are the most prominent features, giving the land the fluency and progress it otherwise lacks. If the map needs to arrest all movement in the land, freezing it in time and representational stasis, it reinvests in the dynamics of landscape by foregrounding rivers as the pulsating 'veins' of the country. The lines for rivers, the coastline, and the natural landscape compete graphically with the lines for administrative borders: while the map needs to show the land as little more than a flat and undifferentiated expanse of ground, it can make definite statements about the political and proprietary claims inscribed on its surface. Turning from *Anglia* to the next page, the map of Cornwall,[21] the difference in scale requires us to adjust our vision; as our eye 'zooms in' on landscape details we move from the lofty national summit to the ground level of the individual counties. Horizontally, their sum adds up to the nation at large; vertically, they refer us back to the opening image. As one continues to turn the pages of the collection the maps introduce the whole range of English topography, concentrating on the toponymic particularity of the area confined by the limits of a county but ignoring almost completely what lies beyond its borders. There is on each map a quickening of interest around the immediate centre which disappears into a shadowy indistinction at the margins of the image. Some maps depict a single county, others crowd several onto the same sheet. All follow the pattern of showing north at the top, while the scale of representation needs to be constantly adapted to enable the structural coherence of the atlas format. This shifting of scales produces undesired effects. Small scale results in a great density of visual detail, creating a strong sense of topographical unity. Large scale, however, easily creates the impression of a 'dissolving' landscape, the gradual loss of an overall territorial coherence. Foregrounding not the relationship between places but the empty space between cartographic symbols, Saxton's maps frequently suggest not an organic network but a bewildering lack of cartographic consistency.

The national theme of Saxton's atlas elicited enthusiastic responses from contemporaries. According to George Owen the maps were 'daiely pervsed by [all Noblemen and gentlemen] for the better instruction of the estate of this Realme, especiallye touchinge the quantitie, scituacion, forme and speciall places of note of all the sheeres of this Realme'.[22] Highly popular among the gentry, Saxton's maps evidently succeeded in pressing the national landscape into a standardized pictorial code. The uniformity of the atlas format, its systematic representational pattern, subsumes all topographical difference under a structural unity and the reading of Saxton's maps becomes an exercise in practical semiotics, a constant act of 'weaving infinite variety into a unified discourse'.[23] Thus clearly hovering between the antithetical poles of sameness and difference, it is perhaps surprising that Saxton's cartographic construction of national space is most frequently seen as a visual and textual exercise in portraying only the latter – that is, the variety of topographical semantics. Morgan, in an essay evaluating their relevance as tools of government, claims that Saxton's project was the mapping of 'the internal morphology of England, both topographical and political'.[24] This view echoes contemporary attitudes. John Gregory, writing about Saxton's 1583 wall-map[25] (but including the county maps in his praise), described it as 'exact and useful' and considered it to be so comprehensive 'that the smallest Village may bee turn'd to there; *Henxey* or *Botlie*, as well as *Oxford*.'[26] Many similar appraisals, both contemporary and modern, could easily be quoted.

But wealth of cartographic detail is not what everybody found, or continues to find, in Saxton's maps. In 1625 John Norden remarked about his own, partly Saxton-based maps in his English travel guide that 'it is not possible for a stranger in so many parts of the Kingdome, to be so well acquainted with Townes and Parishes, as to be able to distinguish the worthiest, *seing they are of like impression in the Maps*'.[27] This reminder of the cartographic erasure of spatial difference recalls the point made recently by Denis Wood that maps are always selective visions of space, graphic evidence of a '*prior* editing' of the world: '*before we get to the map* the domain of expected detail has been throttled down to a certain class of objects.'[28] An inventory of the symbols used by Saxton would have no more than five relevant entries – the signs for hills, rivers, settlements of various forms, enclosed parks and trees. Within their groups, as Norden was clearly aware, none of these symbols allow much internal differentiation. The national image Saxton's maps create is almost devoid of detail, consisting largely of gaps and vacant land – and thus hardly (aesthetically at least) a mirror of England's 'internal morphology'. In a sense, these maps do not chart regional idiosyncracy but a levelling sameness, a repetitive similarity. Looking at them in sequence tells you little beyond the fact that the land of England and Wales is the same wherever you happen to be.

To some extent the concept of the atlas must produce precisely that effect: it is the cartographer's objective to transfer any space into the same set of representational parameters in order to enable the mental operation of reading a known landscape into the abstract map design; it is also this pictorial strategy which most directly echoes the rhetoric of national unity. And further, as Gregory reminds us, calling Saxton's collection an exercise in graphic minimalism hardly does justice to its contemporary appreciation. Of course, the large wall-map to which Gregory referred differs from the county maps in foregrounding the figure of the whole rather than the parts of a jigsaw. It not only accommodated more information but also reassembled the individual maps of the atlas into one coherent statement. But Gregory's impression of an abundance of detail may point to quite a different aspect of Saxton's map than to the illusion of semantic richness. In emphasizing its navigational value as an 'exact and useful' reference work, where any village 'may bee turn'd to', Gregory implicitly defines the map as a visual inventory of local toponyms, a pictorial dictionary of English place-names listing even the most insignificant hamlet. If the map is seen as no more than a spatialized diagram, stretching across a picture plane instead of filling the pages of a book, Gregory's praise of cartographic detail may concern less the attention to particularity itself than the value of the frame enabling this comprehensive vision of space; he reads the form of the map, rather than its specific content, as a significant achievement in classifying disparate geographic data.

Of course, Gregory's comments may not really be about Saxton at all but about what his name had come to stand for. At the time he was writing Saxton's maps had long been supplemented by other cartographic collections, particularly the post-1607 editions of Camden's *Britannia* and Speed's *Theatre* of 1611. Central to both these works was a feature wholly uncharacteristic of Saxton's county maps, the explicit combination of historical text and cartographic image. It was only in his 1583 wall-map that Saxton began, and then only hesitantly, to embellish the image of the land with additional information that placed the land in the discursive contexts of history, law, and chorography.[29] To establish these links on a map was not just a question of aesthetic preference. In 1621, Peter Heylyn claimed:

As Geographie without Historie hath life and motion but at randome, and vnstable: so Historie without Geographie like a dead carkasse hath neither life nor motion at all, and as the exact notice of the place addeth a satisfactorie delight to the action: so the mention of the action beautifieth the notice of the place.[30]

In mutual isolation, both motionless history and lifeless geography resemble a 'dead carkasse', an image reminiscent of Borges' rotting map of the empire. For Heylyn, the dynamism of topographical representation could only result from a fusion of geography and history, from the simultaneity of the visual and the verbal. 'Mappes', Camden noted, 'allure the eies by pleasant portraiture' but only 'when the *light of learning* is adioined to the speechlesse delineations.'[31] The attempt to textually define the visual image of English topography, to impose on the land the 'controlling linearity of narrative description'[32] and thus avoid the multiplicity of political readings to which Saxton's county maps were ultimately liable, responded directly to the agenda of the Tudor antiquarian project that Speed and Camden, but not Saxton, wholeheartedly embraced. Although Saxton's maps were also linked to a chorographical narrative, William Harrison's *Historicall Description of the Islande of Britayne* (1577), the opening text of *Holinshed's Chronicles*, this link was neither explicit enough to have text and map assembled in the same publication, nor did Harrison's topographical account offer the historical dimension Saxton's maps were ostensibly lacking.[33] As the collection stands it has 'life and motion but at randome' and it was not until the publication of John Speed's atlas in 1611 that 'the mention of the action' explicitly beautified – and hence gave meaning to – 'the notice of the place'.

John Speed's popular *Theatre of the Empire of Great Britain*, a regional atlas of England, Wales, Scotland and Ireland, was primarily intended as an adjunct to Speed's history of Great Britain.[34] The programmatic title announces a conceptual redirection. The atlas is dedicated to James, the 'Inlarger and Vniter of the British Empire',[35] whose attempt to achieve the political union of the three kingdoms developed into a highly controversial issue of his reign. The map that opens the *Theatre* (Plate 4) reflects this political agenda: significantly expanding Saxton's *Anglia* it covers the entire area of the British Isles and pictorially creates the impression of inner union Nowell's *General Description* so visibly lacks. Ireland no longer presents an obstacle to the cartographic effect of spatial cohesion. This is most pointedly suggested by the way the Kintyre peninsula visually connects with the county of Antrim, acting almost as a landbridge between Britain and Ireland. On Saxton's *Anglia* both islands are clearly divided by several miles at this point, a detail Speed reproduces on his own individual map of England and Wales (Plate 7). This discrepancy is not just a matter of geographical inaccuracy. Speed, unlike Saxton, was not working from original surveys but collated topographical information from a range of different sources. His maps are deliberate aesthetic statements about culturally determined landscapes, not just visual records of accidental geographies. A map of England and Wales deals with the integrity of a single kingdom, and consequently emphasizes its existence as a separate spatial entity; but a map of 'the empire of Great Britain' promotes the concept of an overarching spatial unity of the British Isles, translated into the visual coherence of the cartographic picture plane. The suggestion of a physical link between both islands is a geographical detail that serves to naturalize this political subtext. On Speed's map, the code of the national which produced the space of Saxton's *Anglia* is enforced and expanded to achieve a congruence between cartographic representation and the monarchical claim to supremacy over the entire 'empire of Great Britain'.

The *Theatre* is the summary of four decades of English cartographic activity: most of Speed's maps are based on Saxton's while some acknowledge their debt to other map-makers such as John Norden or William White. Speed covers England and Wales comprehensively in 62 maps, devoting a map to each individual county, while Scotland and Ireland are present with one general map each, supplemented – in the case of Ireland – by four additional maps showing the four historical Irish provinces. Apart from the extension of geographical space from England and Wales to Britain and Ireland, this programme resembles Saxton's closely. The pictorial arrangements on the individual maps, however, are evidence that Speed did not simply copy the work of his predecessor but that he transformed the descriptive groundwork provided by Saxton into a spatial narrative of national autonomy. The presence of a distinct textual progression is already announced by the sequence in which the maps are compiled. Starting in Kent, the southeast corner of his general map, the reader is led along the southern coast to Cornwall, up to Somerset and back across eastward to Norfolk. This spiral continues through the rest of England until all the counties have been systematically covered. This is not an abstract alphabetical arrangement but the visual reconstruction of the perambulation of the surveyor: the maps not only gradually assemble Britain as a whole, but also weave its topography into the textual progression of a concrete itinerary, a structured route.

Most of the maps in the *Theatre* contain a small cartouche with the inscription 'performed by John Speed'. Other cartographers are occasionally acknowledged whose work is 'augmented by John Speed'. This comment is not simply a mild euphemism for plagiarism. Most obviously, it refers to the plethora of additional information, usually surrounded by elaborate ornamental borders, which Speed crowds into the margins of his maps: city views (some of which made their way into the final volume of the *Civitates* collection), inserted smaller maps, cartouches showing the peculiarities of the regional landscape or impressive local buildings, portraits illustrating characteristic local costume, coats of arms with the titles of the local gentry, lists of place-names, illustrated battle-scenes, detailed views of castles or colleges, pictures of Roman coins, textual references to relevant episodes of English regal and military history, explanations of Roman remains such as Hadrian's wall, images of books, globes, instruments, and so on. With the exception of the decorative frame and the royal coat of arms, all of this was absent from Saxton's maps. Seeing that this additional information might all too easily deflect attention away from the geographical content of the images, what is its purpose? The biographical answer would emphasize the cultural milieu from which these maps emerged. Unlike Saxton's, Speed's atlas was a product of antiquarian thought and effort, and as a work that did not receive any sponsorship from a government in search of authoritative representations of the kingdom it catered less for the needs of the crown than for the intellectual preferences of a set of historically minded local scholars.

But the visual language of these maps points to further important changes which four decades and a series of shifts in the cultural landscape have brought about, changes that involve not merely the stylistic code of the antiquarian movement but the very politics of cartographic representation. The pictorial richness of his cartographic images confers on Speed's maps both a partisan political vision of national history and the immediacy of a narrative presence. Producing a detailed network of causal relations across time and space, these maps move beyond the basic 'plotless' structure of Saxton's cartographics. As Helgerson has noted, the antiquarian project that fuelled these changes was not a neutral scholarly exercise. Against its own intentions, perhaps, it promoted the concept of a land-based nation and thus deflected attention from the monarch as the sole focus of national loyalty. In Speed's atlas this process is most clearly visible in the elaborate heraldic presence of the gentry which marginalizes the royal coat of arms and breaks down the hierarchical concentration on the insignia of the ruler. Despite its explicitly royal agenda Speed's cartographic imagery thus contributed to the emergence of a conception of England defined not exclusively in relation to its monarch, but to all the leading families of the gentry.[36] But heraldic symbols belong to only one specific signifying system at work on the map surface – 'nonlinguistic, conventional'[37] – which principally addresses the issue of political authority. Within the frame of each image the coats of arms compete for attention with a host of other images – portraits, historical scenes, objects of commercial or scientific interest, and so on – which belong to the 'mimetic and iconic'[38] signifying system that conceives of space not only in terms of political power but in terms of its social, historical and cultural meaning. A third signifying system, entirely linguistic, begins with the toponymic structure on the map surface and continues beyond the image on the verso of the printed sheet, where each map is supplemented with a two-page chorographical description. These articulate frames not only redefine the link between king and country along the lines Helgerson suggests, they also change Saxton's silent maps into dynamic cultural narratives, giving them the 'life' and 'motion' Heylyn was later to demand.

The comparison between Saxton's *Anglia* and Speed's *Kingdome of England* (Plates 6 and 7), one of the few maps in which Saxton is acknowledged as a source, demonstrates this point. Hardly any difference in cartographic shape is noticeable and the geographic referent is identical: an England that has fully absorbed Wales. Speed shows a higher density of toponyms and privileges man-made over natural landscape elements; he also extends the list of 52 counties to include statistical material on the number of cities, market towns, rivers and bridges. For obvious reasons, he omits the arms of Thomas Seckford, Saxton's patron. But in terms of geographical information neither map represents any significant advance over the other. The crucial pictorial difference between the two maps is the inclusion of eight portraits, four on either side of the image: a 'lady' and a 'noble man', a 'gentleman' and 'gentle woman', a 'citizen' and his 'wife', as well as a 'countryman' and a 'country woman', nestle in the margins, gathered around the land they inhabit. These portraits are types, not individuals – they represent the inner hierarchy of contemporary English society. The king, it is interesting to note, is displaced onto the map of Scotland, the country of his birth, but here on the map of England nobility and gentry, country and city are all represented. England is their space, their text. As if to rectify a general representational problem endemic to maps – where landscape, almost by definition, is empty and depopulated – Speed implicitly follows Cyprian Lucar's recommendation 'to write in the sides or bottome of the map . . . the disposition, industrie, studies, manners, trades, occupations, honestie, humanitie, hospitalitie, apparell, and other morall vertues of the inhabitants of the described land'.[39] The pictorial map, displaying people in relation to their land, developed into a particularly popular genre in the seventeenth century.[40] On Speed's *Kingdome of England* these portraits turn English topography into a social script that gives the land a voice and a narrative structure: though the narrow scope of these eight portraits suggests a rigid and intransigent society – strict hierarchies, a natural order of heterosexual couples,[41] limited social mobility – the introduction of a 'popular' element into the map design constructs the prosopopoeic image of a national space performed by its 'users' and shaped by the continuous, daily effort of its loyal inhabitants drawn from various social ranks, belonging to urban or rural spheres, but all mythically united in the vanishing point of the national.

Maps, according to Jurij Lotman, belong to the genre of 'classificatory (plotless) text[s]' that denote in terms of language a range of given elements belonging to the world they set out to describe, 'insist[ing] on a definite order of internal organization.' Such texts, Lotman argues, may always assume secondary layers which superimpose a plot, a meaningful chain of events, on the 'basic plotless structure'. Recognition of specific plots depends on the cultural context in which they circulate, and the discursive communities they address. On maps, a plot may take the form of a simple line drawn across its surface, indicating the route of a journey and thus introducing an action which already 'surmounts the [geographical] structure'.[42] This is precisely the function of the portraits included on Speed's *Kingdome of England*. They define the nation not as a static descriptive figure but as a dynamic historical agent. On the maps of the *Theatre* the spatial continually intersects with the social and temporal, not just in the combination of atlas and history but within the pictorial limits of the map itself. With the help of the mimetic icons spread over the surface of the maps, the representation of an autonomous land is linked to a recognizable set of historical personae moving through the semi-fictional space of the national theatre, giving it temporal depth and a distinct cultural profile.[43] This newly territorialized 'plot' of the nation is more than merely self-congratulatory in intent: it defines England as a negotiable public sphere beyond the limits of the royal body politic, but simultaneously introduces new limitations by aiming to standardize – through its use of generic portraits, its affirmation of an 'official' past – the social and political configuration of the emerging nation. After all, as embellishments of the space they inhabit these figures suggest that the members of the national community, as well as the land they signify, are equally fit objects for the geographical 'dissection' performed by the cartographer.

This fusion of word and image, notably absent from Saxton's conception of national geography, is an essential element of Speed's cartographics. Grounding the map deeply in historical time, he extracts from national space a textual narrative that explicitly links the 'British' territory – as Camden and Drayton also did – to the uniqueness of its historical, poetic and antiquarian wealth. This enables him to 'emplot' the antiquarian notion of Britain in the image of a gentrified nation, firmly inscribing, as Lotman's model suggests, a secondary layer of meaning on Saxton's classificatory text, or 'plotless' structure. On Speed's maps, the gesture of cartographic emplotment, grafting the verbal onto the visual, thus claims the nation as a space of emergent self-recognition. If national history has to be told and retold before it takes root in social memory, and if the space visually displayed on a map needs to be strictly temporalized in its service, then Saxton's maps lack the narrative dimension constitutive of the national idea:[44] a route, an itinerary, a specific narrative progression may only be speculatively extracted from the ensemble of landscape elements they offer, and in each concrete instance these routes are not realized but articulated only as the abstract sum of their possibilities.[45]

The discussion in this chapter of three distinct moments in the early modern mapping of Britain follows Harley and Zandvliet's suggestion that if 'the emblems that qualify and frame the maps are part of an ideological dialogue, then it is more probable that the geography itself is discursively embedded within broader contexts of social action and power.'[46] Heraldic symbols, references to patrons or representations of the monarch constitute not just ornamental additions to maps and atlases but subtexts of power and authority which organize the cartographic image as specific cultural and political statements. But in the same manner that the frame reflects the external contexts in which the map circulates, its geographical content is subject to the discursive pressures that enabled its production. Imagining the cultural and geographical boundaries of the national community were tasks that crucially overlapped for early modern English cartographers and the choice of what to include and exclude on their maps was not merely a pragmatic decision. The signifiers used by Nowell, Saxton and Speed, within less than five decades of each other, for the political space their maps describe – 'our region', 'Anglia', 'the empire of Great Britain' – are evidence that cartographic images are not neutral mirrors of spatial facts, existing outside and beyond their representation, but textual and pictorial sites on which the changing vision of a national imagination found its articulation. Thus, if a rhetoric of nationhood presupposes a geographical discourse that enables the definition of frontiers and boundaries,[47] cartography constitutes the medium in which not only the geographical, but also the social and cultural referent of the emergent nation was given both shape and meaning.

6
The Image of Ireland

Compared with the popularity of early modern maps of England, the cartographic shaping of Ireland seems to have developed at a much slower pace. Although a good governor in Elizabeth's administration, as Robert Beale advised in 1592, needed to sharpen his geographical knowledge with 'the booke of Ortelius' Mapps, a booke of the Mappes of England . . . and also a good descripcion of the Realm of Irelande',[1] no description of Britain's neighbouring isle existed in print that could even remotely rival the visual impact of Ortelius' or Saxton's atlas. It was not until 1685, more than a century after Saxton's maps had started coming off the presses in England, that William Petty published his *Hiberniae Delineatio*, the first complete Irish regional atlas. Petty had fully surveyed the country almost 30 years earlier, in the wake of the Cromwellian reconquest of Ireland, but even 1659 – the year he finished the *Down Survey*[2] – seems remarkably late when contrasted with the advanced state of English regional cartography. This delay has attracted a range of rather unconvincing explanations – such as the difficulties and dangers of travelling, bad weather conditions from which Irish field-workers were more likely to suffer than their colleagues in England, or the occurrence of, in J. H. Andrews' euphemism, 'expressions of antagonism by the local population'.[3] As we have seen, popular dislike of the surveyor's presence in the countryside was a common feature in England. In Ireland there prevailed an even stronger sense of equating map-making with a hostile intrusion into an ancient landscape. Surveying was considered a form of territorial invasion, designed to dispossess people of their land, which called for rigorous resistance – and at least one contemporary cartographer, Richard Bartlett, fell victim to such harsh attitudes when he was decapitated shortly before finishing his survey of Donegal.[4]

The 'piecemeal, *ad hoc* fashion in which the mapmakers were put to work'[5] in Ireland might of course simply suggest that the English government never considered the compilation of a systematic, county-by-county atlas of Ireland a matter of particular urgency. Beale, whose further recommendation that 'if anie other plotts or mapps come to his [the secretary's] hands, let them be kept safelie'[6] has particular resonance with regard to Ireland, employs a significant threefold distinction: for the English administrator Ortelius' world atlas needs to be backed up by two works of chorography – a regional atlas of England (presumably Saxton's), but only a mere 'descripcion' of Ireland, an expression which seems to suggest the less pressing need for a thorough acquaintance with the Irish territory. Yet to imply a lack of cartographic interest in Ireland's topography on the part of the Tudor government would be to seriously misrepresent the facts, a distortion we may be liable to on account of both the decorative appeal of the printed atlas, and its status (in England) as an emblem of the collective idea of nationhood. In Ireland, precise territorial knowledge was considered sensitive political information and one would not expect such data to be freely available. Irish maps were predominantly strategic responses to pragmatic needs, designed for highly functional contexts, and not intended to celebrate the political community that claimed ownership of both map and territory. Although some highly decorative maps survive (some examples are discussed further below), most of the maps drawn in this period and later were either commissioned directly for military purposes – since in Ireland, much as in other parts of the world, the cartographer arrived in the shadow of the soldier – or catered for legal needs – depicting, for instance, the confiscated lands to be distributed among newly recruited settlers.[7]

Government officials in Ireland, part of the prospective readership of Beale's tract, would have had access to manuscript material, of course, and as the library of Lord Burghley testifies, maps of various kinds have always been of great importance to successive Tudor administrations of the country.[8] Indeed, a look at the cartographic evidence other than the decorative atlas shows that the map-makers' interest in Ireland even outscored cartographic activities in England during the final phase of the Tudor conquest – that is, the period 1580–1603 – when 'more maps were drawn of [Ireland] and of its provinces and districts than of English regions.'[9] The abundance of sixteenth-century Irish maps – today scattered over numerous archives – is evidence that the diligent if unsystematic surveying and mapping of Ireland, especially of those parts of the country selected for the government's plantation schemes, had been a continual process from at least the mid-century onwards.[10] Despite the absence of a comprehensive sixteenth-century Irish atlas, therefore, Ireland hardly played a marginal role in the geographical awareness of the English government. But the impression of a lack of Irish maps unwittingly repeats a contemporary theme, if only by way of analogy. Perceived as culturally distant and physically inaccessible, both by government officials and professional cartographers, Ireland continued to elude English military and discursive control in Elizabethan times. Irish maps reflect this incomplete conquest: even where an uninterrupted line encloses the entire island, suggesting full cognitive possession of the space it circumscribes, the visual and conceptual coherence of the pictorial surface is never fully achieved.

The present chapter argues that Irish maps are located at the interface of contrasting political desires, a point I will substantiate by looking in turn at various general maps of Ireland, a set of military campaign maps, and finally a plantation survey. This selection quite deliberately cuts across traditional principles of classification. These maps all belong to different genres, address different audiences and serve different goals. In particular, the selection disregards the distinction between the popular appeal of the colourful wall map or the printed atlas, and the administrative, legal and military needs met by large-scale regional maps, which frequently remained in manuscript. But common to all cartographic activity was a general 'mapping impulse'[11] which links maps of various genres to a set of mutually shared interests and invites us to examine all maps of Ireland as products of specific Anglo-Irish negotiations of space. I noted in Chapter 3 how John Hooker's mention of the arrival in Ireland of surveyor Robert Lythe indicates that English maps of Ireland reflect wider ideological investments in the land than a mere analysis of their geographical 'accuracy' would reveal. Rather, these maps are graphic evidence of the contradictory processes by which Ireland's landscape may either be visually absorbed into the emerging political concept of national British unity, marked as a locus of cultural difference, or even ignored completely as a geographic insignificance located safely beyond the borders of the British mainland. In other words, maps of Ireland should be discussed, as I propose to do now, not merely as containers of geographic data but as pictorial sites of cultural and political conflict.

A convenient place to begin this discussion is the Irish section of Laurence Nowell's *General Description of England and Ireland* (1564/5, Plate 5), a map I have already briefly looked at in the previous chapter.

As we have seen, this map includes among its ornamental additions the portraits of two male figures, one in each of the bottom corners. The reclining figure on the left, visibly distressed and attacked by a baying hound, leans on a pedestal bearing a Greek inscription from Hesiod's *Works and Days* stating that after Pandora's opening of the jar '[h]ope was the only spirit that stayed there / in the unbreakable / closure of the jar, under its rim, / and could not fly forth'.[12] The seated figure on the right, haughty and demanding in appearance, has a few books scattered about his feet and rests on an hourglass underneath which another inscription from Hesiod advises to 'hope and endure'.[13] As I noted above, these figures are usually identified as straightforward portraits of Nowell (left) and Cecil (right); and they are included on the map, it is argued, as a biographical footnote on the strain of a patronage relationship. They are thus read as part of the presentational frame of the map, not as properties of the cartographic image itself. The biographical identification, though speculative, is hardly controversial, but to enquire no further into the meaning of these portraits means to ignore the semantic richness of a complex pictorial arrangement. For if the portraits are read as a 'scene' cut off from the geographical information, the map image is divided up into several layers of signification, essentially unrelated to each other. If, however, the map is viewed as an integral whole of which the figures form only one, albeit a vital, part, one immediately recognizable set of relationships, one that determines not just a section of the map but the entire picture plane, links the cartographer and his patron to the images in the top corners, the title cartouche and the coat of arms. As a pictorial ensemble these four images thus form a comprehensive framing device, turning the land into an object of both royal ownership and authorial intention.

Read as allegories for the map's political and geographical content level, the portraits are even more suggestive. Representing the two halves of an unequal political union between a superior England on the right and a dejected Ireland on the left, their respective attitudes are indicative of the political relationship between Britain's 'core region' and its most recalcitrant outlying province, the stubbornly rebellious Ireland. In so far as both figures assume classic postures of melancholy, they amplify the English–Irish duality into a double state of dejection;[14] but it is the depressed stance of the cartographer, attacked by the bane of surveyors, a baying hound, which condenses into one image the English frustration with the canine Irish, perceived as savage and barbaric, over whose territory the seditious contents of Pandora's box hold sway at will. The cartographer's empty purse is as much an illustration of this mythological subtext as an allusion to the poor state of Nowell's finances and the squandered resources of Ireland. Thus, in its function as an address to the culture that produced it, Nowell's map does not reflect an 'objective' sixteenth-century perception of British and Irish geography but constructs an image of the internal dynamics of a political space which obliquely, through the use of textual and visual ornamentation, acknowledges cultural difference as a defining characteristic of the Tudor state.

English attitudes towards Ireland in the sixteenth century were increasingly marked by the contradiction between the insistence on absolute Irish otherness, rationalized as a fundamental incivility, and the need to assimilate the island into some sort of national framework on account of the political danger resulting from its geographical proximity. As Michael Neill has recently noted, '[w]hile the ideology of national difference required that the Irish be kept at a distance and stigmatized as a barbaric Other, the practicalities of English policy more and more pressingly required that Ireland be absorbed within the boundaries of the nation-state.'[15] This conflict is registered on Nowell's map in the cartographic contours of Britain and Ireland which are implicated, to borrow a phrase from John Gillies, in a '*semiosis* of desire'.[16] On the first printed map of the British Isles by George Lily, published 1548 in Rome (Plate 8), Britain still owes its upright shape partly to the medieval tradition of the 'Gough' map. Flanked on its western shore by an oddly shaped smaller island that does not really seem to belong there, it adheres to a perfectly straight line from south to north and stands erect if the map (which has south to the left) is turned to suit our viewing conventions. But Nowell's Britain visibly curves westward as if to bend over and encircle its neighbouring isle. The political statement of the map – the description of a space that aspires to the collective vision of a fully anglicized terrain – is thus translated into an almost physical incorporation of those areas which, viewed from the English centre of power, constitute the outlying regions of an unevenly structured polity, the culturally diverse margins of the Tudor state. Yet the image of Ireland disputes this scenario by visually resisting such geographical appropriation. Large clusters of green suggest the intractability of a wild and barbaric landscape, its rough texture spells out the absence of spatial order. Pulled westward by the dynamism of the cartographic shape the viewer's gaze centres not on the dense toponymic surface of Britain but on the graphic irregularities and textual gaps of an 'unfinished' Ireland, acting as the constant reminder of the incom-

1 Johann Vermeer, *The Geographer* (1668). Städelsches Kunstinstitut Frankfurt am Main.

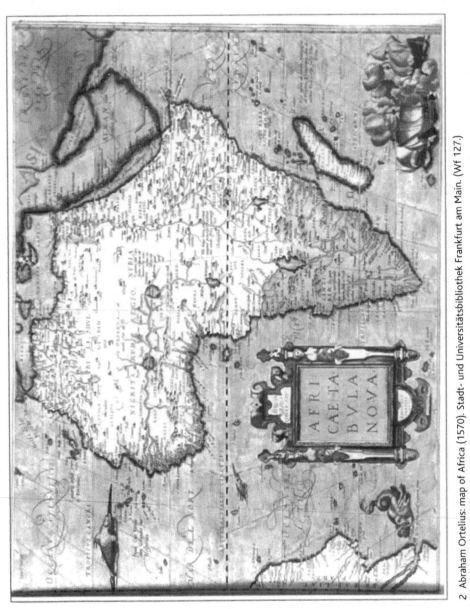

2 Abraham Ortelius: map of Africa (1570). Stadt- und Universitätsbibliothek Frankfurt am Main. (Wf 127.)

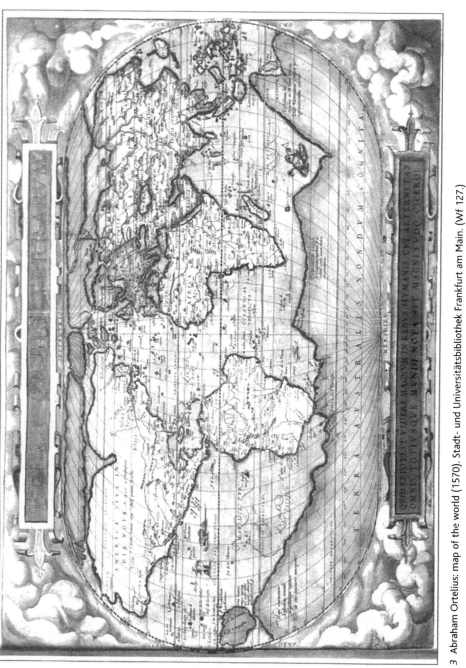

3 Abraham Ortelius: map of the world (1570). Stadt- und Universitätsbibliothek Frankfurt am Main. (Wf 127.)

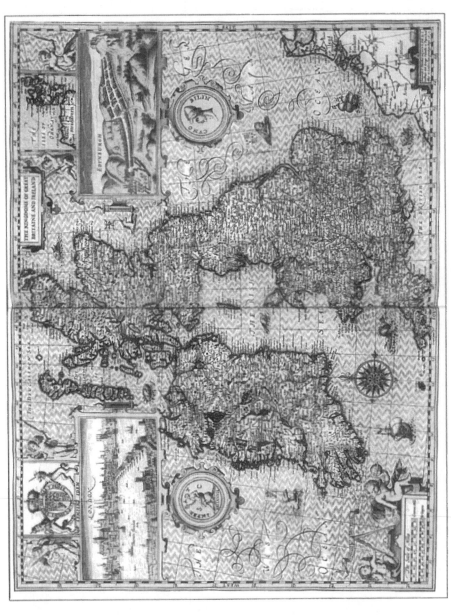

4 John Speed, *The Kingdome of Great Britaine and Ireland* (1611). By permission of the British Library. (Maps C.7.c.20.)

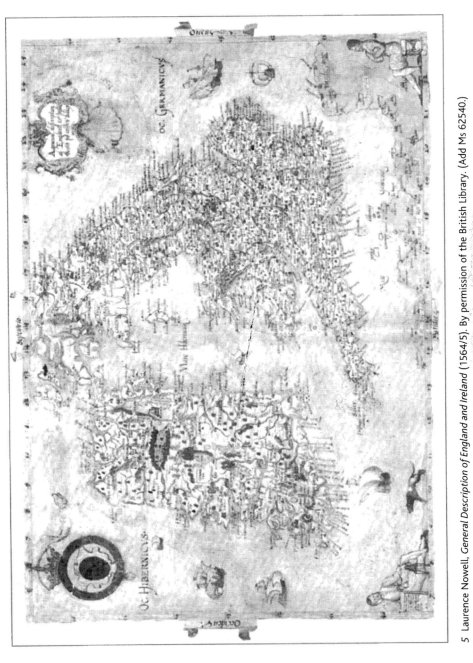

5 Laurence Nowell, *General Description of England and Ireland* (1564/5). By permission of the British Library. (Add Ms 62540.)

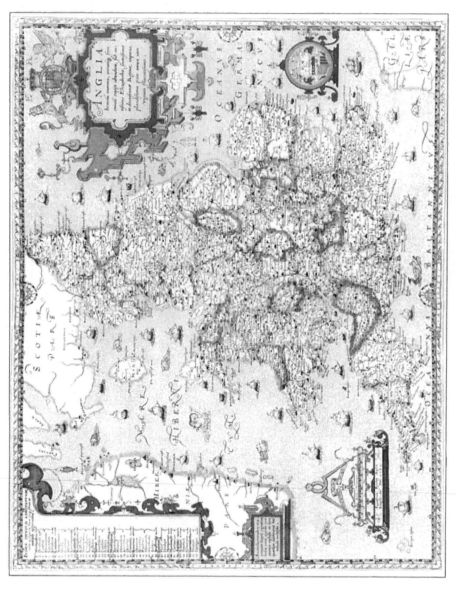

6 Christopher Saxton, *Anglia* (1579). By permission of the British Library. (Maps C.7.c.1)

7 John Speed, *Kingdome of England* (1611). By permission of the British Library (Maps C.7.c.20.)

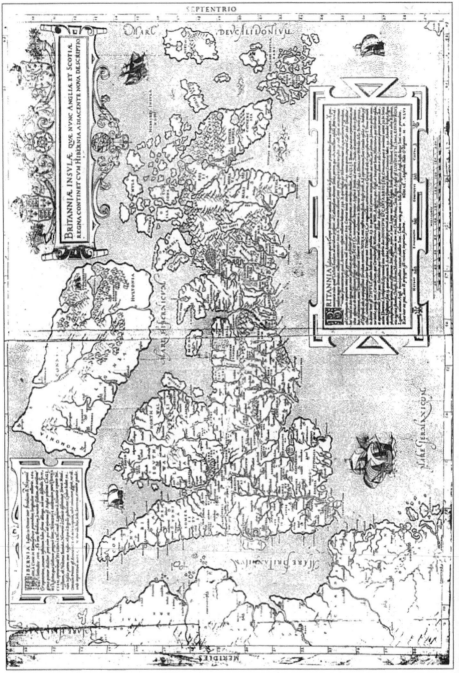

8 George Lily, *Britanniae Insulae* (1546). By permission of the British Library. (Maps K.Top.5.1.)

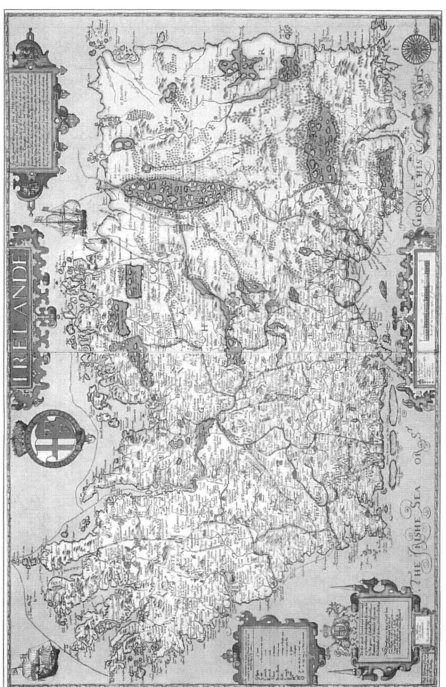

9 Baptista Boazio, *Irelande* (1599). By permission of the British Library. (Maps C.2.cc.1.)

10 John Goghe, *Hibernia* (1567). Public Record Office, Kew. (MPF 68.)

11 John Speed, *Kingdome of Irland* (1611). By permission of the British Library. (Maps C.7.c.20).

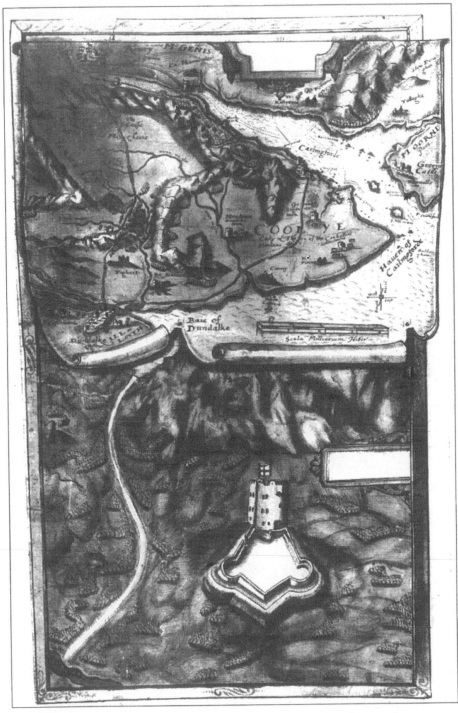

12 Richard Bartlett: map of Moyry Pass (1602/3). National Library of Ireland. (Ms. 2656.)

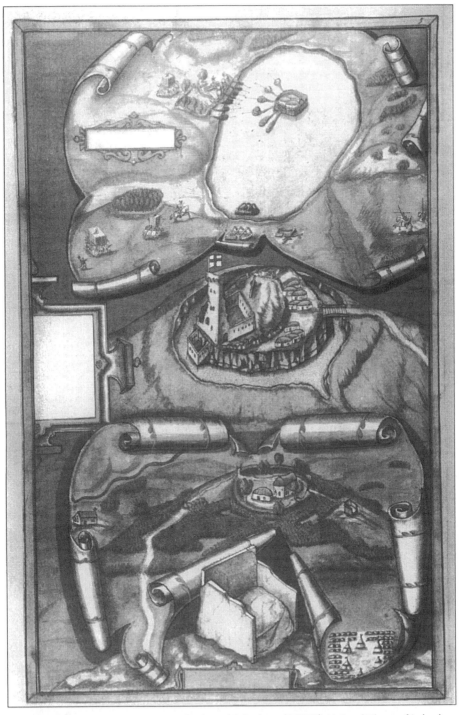

13 Richard Bartlett: map of Dungannon Castle and Tullaghoge (1602/3). National Library of Ireland. (Ms. 2656.)

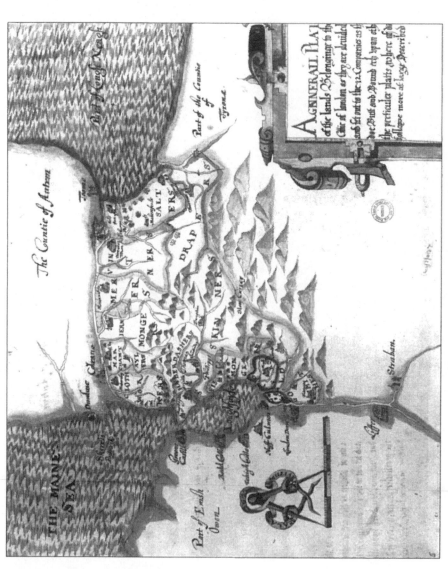

14 Thomas Raven: the county of Londonderry (1622). Public Record Office Northern Ireland (T/510/1.)

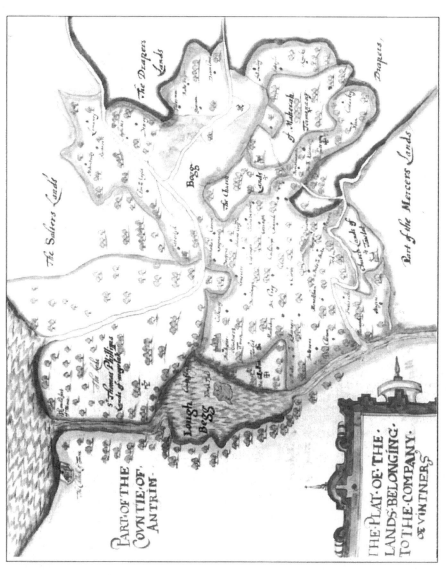

15 Thomas Raven: the Vintners' lands (1622). Public Record Office Northern Ireland. (T/510/1.)

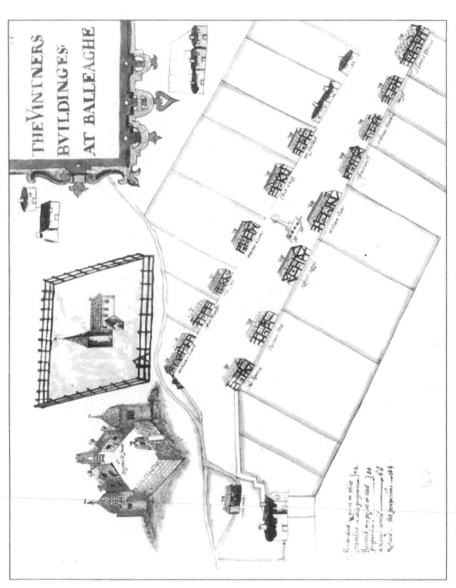

16 Thomas Raven: the Vintners' buildings at Balleaghe (1622). Public Record Office Northern Ireland. (T/510/1.)

plete conquest. The map's temporal code – the *memento mori* of the hourglass, the plea to 'hope and endure' – further supports this reading by consigning to the future the uncertain political and cultural unity of the area Nowell describes, in his letter to Burghley discussed above, as 'regionem nostram' – *our region*.[17]

Despite Burghley's initial interest Nowell's map never reached print and it was not until the end of the century that the first single map of Ireland came off an English press. This was Baptista Boazio's *Irelande* (Plate 9), a highly decorative map which owes much to earlier efforts, particularly to Mercator's version printed four years previously. Now commonly dated to 1599, the map has recently not met with much approval on the part of historians of cartography. Undoubtedly an example of accomplished craftsmanship, its lavish ornamental flourish, the purely fictional character of some of the map's topographical details and the way Ireland is visibly 'encased' by the decorative frame suggest that precise geographical information was not the map's principal objective. Both Boazio, the cartographer, and Elstrack, his engraver, are doubly present on the map: while two cartouches bearing their names signal directly their claim to authorship, a more imaginative but deeply colonial gesture transforms them into the toponyms *Baptiste's Rock* (off the Antrim Coast) and *Elstrake's Isle* (south-west of Tyrconnell). This inventive way of writing cartographer and engraver into Ireland's geography has led one commentator to suggest that the map 'is not a good one, even by contemporary standards: obsolete before it was published . . . its geographical content is badly garbled and in places totally fictitious.'[18] Such a view, though factually correct, implicitly assumes that the gradual increase of cartographic accuracy should be seen as the guiding principle of map history. In my view, what makes Boazio's map such an important example of the way sixteenth-century Englishmen made spatial sense of the intractable and 'barbarous' Irish territory is precisely its value as a decorative image of Ireland's geography fluctuating between fact and fiction. Its purpose was not accuracy but opulent display. Boazio's and Elstrack's names function as a kind of geographical signature, an eccentric gesture perhaps,[19] but one that capitalizes on Ireland's status as the property of those that give it visual and verbal presence in maps and texts.

Nowell's Ireland, as I have argued above, oscillates between its status as alien other and as an integral part of the national landscape. Cartographically represented as England's object of desire, it eludes English discursive control, causing in response the resigned posture of 'hope and endurance'. Three decades later Boazio's image suggests a hardening of attitudes but still retains marks of these anxieties. The surface of the map forcefully articulates the claim to English supremacy over Irish soil: lavishly spread out over the entire canvas, Ireland is surrounded by symbols of English domination – St George's flag at the top, two majestic English ships sailing the Irish coastline, a dedicatory address to the queen in the bottom left-hand corner crowned by the royal coat of arms. In this dedication, the 'loyall' cartographer hands both land and map to the English monarch, inviting her – in a reverential gesture reminiscent of Blundeville's exuberant praise quoted above – to 'distinckly see' the whole island with all its 'Hauens, Rockes, sandes [and] Townes'. Additionally, almost the entire country is densely covered with English place-names and the names of the larger landowning families. In the left-hand margin we even find a brief English–Gaelic dictionary helping us to decipher Ireland's foreign-sounding place-names. By having successfully transferred the particular landscape of Ireland into an image implying an English conception of land ownership and using a standardized representational code, local knowledge no longer seems to matter: via its cartographic representation the impenetrable Irish landscape has become readable, the terrain '[removed] from the cognitive ownership of those who inhabit it.'[20]

In terms of the map's visual language, however, this act of appropriation is incomplete: north-west Ulster – disproportional, oversized and equipped with a largely fictional coastline – is still predominantly a toponymic void, lacking the proprietary tags spread over the rest of the map. These silences and empty spaces in the midst of what is otherwise a 'garrulous' cartographic surface signify more than just technical difficulties in procuring reliable topographical data. Rather, the shape and texture of this unscripted and inflated corner of the map are the result of anxieties about Irish 'savagery' projected into the physical geography. An earlier Irish map gives further substance to this claim, a map closer to Nowell than to Boazio: John Goghe's *Hibernia* of 1567 (Plate 10). On this map, in the same north-west corner of Ulster, three figures in full military gear visibly guard the terrain, preventing both physical and visual access to the land. These are depictions of Irish warriors, the notorious gallowglasses, who keep the terrain beyond English reach. As a pictorial memento of active native resistance – which literally disappears from view on Boazio's map – the inclusion of these gallowglasses both betrays English anxieties about this region and acknowledges military conquest as a necessary precondition for Ireland's full anglicization, a political vision eventually given full cartographic articulation in

John Speed's Jacobean atlas, *The Theatre of the Empire of Great Britain* (1611).

Speed's 'atlas', of course, is still generically a 'theatre', and before discussing his Irish map more fully below I would like to briefly consider the interrelation of dramatic and cartographic imagery by looking at a set of references to Ireland in four plays by Shakespeare. The point of this digression is to show how both maps and plays, when turning to the matter of Ireland, invest to different degrees in the image of a nebulous territory beyond the porous borders of the national sphere: just as Ireland is rarely more than a shadowy presence on Tudor maps, always only partially coming into view, it resists full spatial appropriation on the Elizabethan stage. *The Comedy of Errors* opens the set of spatial references to Ireland in the Shakespearean canon. In Act 3, a comic exchange between Dromio and Antipholus of Syracuse is structured around the image of a terrestial globe. Prompted by Antipholus, Dromio practices his misogyny on the kitchen maid Nell: 'She is spherical, like a globe. I could find out countries in her.' 'In what part of her body stands Ireland?' 'Marry, sir, in her buttocks. I found it out by the bogs' (III.ii.113–17).[21] Dromio proceeds to allocate different countries to parts of Nell's body: he discovers Scotland in the barren palm of her hand; France in her warlike forehead; England in the chalky cliffs of her teeth; Spain in her hot breath; the Indies in the carbuncles on her nose. 'Where stood Belgia, the Netherlands?' 'Oh sir, I did not look so low' (136–7). Dromio's spatial fiction is enabled by the grotesque female body of Nell, serving as the external framework for a political geography of national stereotypes. Conflated with her body is the metaphorical paradigm of the globe, and by extension the map, inscribing a corporal topography with political and ethnographic meaning. Ireland is singled out for the only explicitly scatological reference – a territorial cesspit and potential source of infection and disease.[22] A bodily hierarchy of high and low, clean and filthy, cultivated and repressed, is translated – via cartographic imagery – into a political relationship of cultural domination. In an Irish context the word 'bog' has an additional resonance that exceeds the allusion to bodily hygiene and points to the wider context of English political anxiety centred on an impenetrable Irish landscape. For Spenser, Irish 'bogs' are dark and threatening hiding-places, shelter of the monstrous Irish kern, 'a flyinge enemye hidinge him self in woodes and bogges'.[23]

In Shakespeare's English histories the symbolic space Ireland occupies in this early comedy gives way to a more direct engagement with its physical reality. In *2 Henry VI*, a messenger arrives on stage with news of an uprising in Ireland.[24] York, who has just been seen plotting against Gloucester, is asked 'to lead a band of men' (III.i.312) into Ireland and crush the rebellion. York complies, but only to use this opportunity for his own plot against Henry. The soldiers in his charge turn into 'sharp weapons in a madman's hands' (347) – once in Ireland, their number will greatly increase through the addition of Irish mercenaries. The Irish rebellion is thus doubly contagious, materially amplifying armed conflict in England and physically spreading across the Irish Sea. The activities of York bring to a fatal climax the interdependence of events in England and Ireland: 'Whiles I in Ireland nourish a mighty band, / I will stir up in England some black storm' (348–9). This black storm will be raised by Jack Cade, 'a headstrong Kentishman' (356), employed to 'make commotion' (358) in England during York's absence. Cade is well trained in Irish techniques of disguise – York, under whom Cade used to serve in Ireland, recounts how he has often seen him spy among the native Irish in the appearance of a 'shag-haired crafty kern' (367). The final act sees York's 'army of Irish' (V.i) briefly invade the stage as a prelude to the English king's defeat in battle, a stage direction that offers a striking parallel between dramatic and cartographic representations of Ireland: on Goghe's map, Irish gallowglasses enter the visual field of the cartographic display; in *2 Henry VI*, they enter the visual field of the Shakespearean stage.

The play presents the Irish intervention in English politics as a highly complex and ambivalent affair. On the one hand, as Andrew Murphy has recently noted, it skilfully recruits Irish otherness in the service of an internal English dispute over royal succession: York is in full control of his 'army of Irish', Cade masters the specific Irish ability to switch identity at random. But although Ireland serves as 'a source of English strength',[25] at least for York, it is also a source of disaster, war and confusion. England's misfortunes are tied up with its involvement in Irish affairs: armed conflict is shown to be endemic to Ireland and spreads to England only through York's intervention; the English civil wars start with an army raised in Ireland that crosses a sea which should have served, according to John of Gaunt, as a defensive wall; and the internal rebellion of Cade – whom John Hooker considered 'an Irishman borne'[26] – is implicitly enabled by his adoption of Irish tactics. Thus, in analogy to the symbolic Ireland of the *Comedy of Errors*, very real dangers are shown to be emanating from its physical space.

In *Richard II*, Irish rebels are again causing political unrest, a situation that necessitates the king's personal intervention. But in contrast to the earlier play, Richard's Irish adventure invites catastrophic failure. Just prior to his return from Ireland his troops desert him and he subsequently loses the entire kingdom to Bolingbroke. Where York could gather military strength, Richard lost it. Ireland's physical territory remains vague and unspecific throughout the play – a distant wilderness, a dark space looming beyond the confines of the dramatic scenery. In contrast to *2 Henry VI*, the Irish are now invisible, they have disappeared as actors, both from the stage and from the scene of English politics. In *Henry V*, written against the background of the very real rebellion of Hugh O'Neill, they reappear in the person of the Irish captain Macmorris. As an Irish soldier Macmorris now serves a legitimate English king, in contrast to the kerns and gallowglasses brought on stage in *2 Henry VI*. During the siege of Harfleur he famously erupts into a string of expletives, levelled at the Welsh captain Fluellen: 'Ish a villain and a bastard and a knave and a rascal' (III.iii.61–2). This outburst in the company of an English, a Scottish and a Welsh captain is framed by the double articulation of what has stubbornly remained, in centuries of Shakespeare criticism, an unanswerable question: 'What ish my nation?' (61/2). By now a principal Shakespearean target of colonial discourse analysis, this brief scene has recently been the subject of various contrapuntal readings which attempt to reclaim Macmorris from his subsumption under the hegemony of English power, insisting that his cryptic remarks destabilize the certainties of the colonial discourse which produced them in the first place.[27] Such critical efforts highlight the crucial ambiguities of this scene but must still concede that Macmorris appears here in explicit support of an internal colonial project that was slowly taking shape when the play was written, the project of a unified Britain. That union is difficult to achieve, as the dramatic row testifies, but the scene nevertheless adumbrates a collective idea of Britishness, a largely fictional political space translated into the schematic quartet of the four captains.[28]

Thus, in Shakespeare, Irish space is initially synonymous with a source of political unrest, a place from which rebellion may at any moment spread to England. It is a physical and political wilderness, a sinister 'secrete skourge',[29] destined to haunt England. In the *Comedy of Errors* Ireland is constructed as a metaphorical territory of human and political waste. In *2 Henry VI* and *Richard II* Ireland meanders between presence and absence, visibility and shadow, and while it may only serve as a historical background to the drama of royal succession, it has a crucial impact on the turn of events. But in *Henry V* Ireland's agency seems notably contained, the 'army of Irish' have been reduced to the character Macmorris and though random aggression still dominates Irish nature it now ultimately contributes to the successful English war effort. Read against the background of the Nine Years' War, Macmorris' irate question about his 'nation' still reveals a fundamental anxiety about the nature of Irish identity and what appears to be an uncanny Irish tendency, personified in the historical character Hugh O'Neill, to change sides whenever a convenient moment arises.[30] But this destabilizing effect is rendered less threatening through Ireland's political domestication, reflected in Macmorris' inclusion in Henry's army. In the political rhetoric of Jacobean England, Ireland increasingly gets assigned a fixed place within a national 'British' framework, an imperial discourse resting on the expanded definition of a national territory encompassing the entire geographical area of the British Isles. Ireland may not be fully assimilated into this framework – the Irish continue to speak deficient English and start unnecessary rows – but the scope for potential transgression inherent in earlier representations is significantly narrowed.

The *Kingdome of Irland* (Plate 11), the Irish map included in Speed's *Theatre*, subscribes to the same national vision as the four captains scene in *Henry V*. Reflecting a new political order, this map enjoyed a far higher distribution than any of its precursors and became the standard representation of Ireland for the first half of the seventeenth century. English military triumph had cleared the ground for the cartographic conquest. In the intervening years since the publication of Boazio's map Hugh O'Neill had been finally defeated by the English general Mountjoy, the successor of the luckless Essex whose ill-fated Irish campaign is alluded to several times in *Henry V*. After decades of intermittent warfare, the whole of Ireland was now under full English military and political control. Accordingly, Speed's atlas can present Ireland as an integral part of the 'empire of Great Britain': if Boazio's map of 1599 only implicitly suggested that colonial Ireland partakes in some vague way of a larger national British territory, this myth of spatial integration becomes the governing statement not only of Speed's Irish map, but of his entire atlas. Graphically, the *Kingdome of Irland* portrays a neat and perfectly controlled area, a peaceful and quiet expanse. A systematic toponymic structure provides the coherent representational pattern absent from all previous Irish maps. Ireland is safely subdivided into 32 counties,[31] implying its spatial subjection to English law, and although the inaccuracies regarding the coastline of north-west Ulster persist, the pictorial surface of the map achieves both homogeneity and balance, suggesting a spatial harmony devoid of conflict. Irish space appears fully synthesized, a flawless unity of landscape. Reminiscent in colour and texture of Speed's own map of England, it is rendered cartographically visible as a perfectly 'natural' extension of 'Great Britain'.

Yet the six portraits Speed places to the left of the map introduce an element of tension that challenges the image of peaceful coexistence. These portraits include the 'gentle man and woman of Ireland', the 'civil Irish man and woman' and the 'wild Irish man and woman' – all wearing the infamous Irish mantle, frequently the object of sharp English criticism.[32] The last entry on this pictorial list defines Irish space most unambiguously as a locus of cultural difference. Disputing the visual harmony of Speed's paper landscape, the 'wild men and women' nestling in the margins of the map are visible signs of Irish 'otherness', occupying the lowest cultural level in the entire *Theatre* and thus offering an oppositional agenda to the dominant plot of tranquil geographical proximity. But they also imply a degree of sameness since on his map of England (Plate 7) Speed employed an analogous device when he included eight portraits that broke down English society into individual social categories – gentry and nobility, citizens and countrypeople. On both images the people are 'mapped' alongside the land they inhabit; in both cases they are separated from the land by being placed in a classic decorative arcade. But these similarities concern the frame of the image, not its contents, and thus only serve to hide the significant switch from the register of social rank employed on the map of England to the divisive and 'proto-anthropological' argument of civility and wildness depicted on the Irish map. To suggest that the similarity of representational features implies an absence of cultural difference within the 'empire of Great Britain' is surely to misread both the specific ethnographic significance of 'wildness' – a self-authenticating device defining not the 'wildness' of others but the 'civility' of the speaker[33] – and the immediate cartographic context of the portraits. Speed can parade the signs of Irish 'otherness' – the mantle, the spear carried by the 'wild man' – because the map ultimately subjects its social script to the discursive control exercised over Ireland's topography: in analogy to the military discipline of Henry's army, which eventually masters even the violent temper of Macmorris, the ordered space of Speed's map does not tolerate the unrestricted movement of 'wild men and women'. The pictorial reminder of Irish barbarism serves not as an image of harmony but as a figure of enforced integration, suggesting the successful containment of Irish savagery.

Viewed in historical sequence, the series of Irish maps discussed in this chapter serves as a sensitive cartographic register of the violent

transfer of political authority in Ireland from native Irish to English colonizers. The 'wild men and women' on Speed's Irish map gesture at the
historical outcome of this political struggle for which the calm and decorative scenario of the map design served as the venue. In contrast to
the openly defiant gallowglasses on Goghe's map, or the invisible inhabitants of Nowell's graphically impenetrable Ireland, the Irish no longer
roam the wild landscape at will but are safely accommodated into a pictorial framework visualizing Ireland's political subjugation. On maps, as
on the stage, Ireland eventually moves from its shadowy location in
threatening geographical proximity to its visible inclusion in the spatial
unity of a larger territorial setting. In an imaginative displacement
Goghe's gallowglasses are forced off the map and wander first into
Boazio's visual void, and then to the margins of the map to be contained in the pictorial frame of Speed's portraits. Similarly, the 'army of
Irish' Shakespeare brings on stage in *2 Henry VI* recede into the shadowy
background of *Richard II*, and finally metamorphose into Macmorris, an
Irish soldier safely contained in another territorial framework, the
'British' army of an English king.

   In pointing out these conceptual affinities between dramatic and
cartographic representations of Ireland I want neither to claim direct
mimetic congruences nor to suggest conscious authorial intentions. The
temporal disjunctions and generic differences between various scenes
written for the London stage during the decade of the 1590s and a series
of maps spanning half a century of cartographic activity from 1564 to
1611 surely do not offer the coherent body of historical evidence a
straightforward comparison would require. But I would like to suggest
that the transmission of the cartographic image of Ireland from Nowell
and Goghe through Boazio to Speed follows the logic of a trajectory
observable also in the fragmentary representation of Ireland in Shakespeare – a logic that increasingly attempts to contain Irish cultural difference in a 'British' framework. Ireland's threatening otherness cannot
be completely erased – its overt and discreet manifestations continue to
be present in all the visual and verbal examples I have so far looked at
in this chapter – but it can be reduced in meaning to an instance of local
peculiarity. The similarity of discursive movements in the representation
of space signified as 'Irish', circumscribing a conceptual triad of menacing presence, forced absence and guarded inclusion, is evidence of a continuous reflection on the link between people and land, pervading
geographical thought, cartographic practice and literary discourse alike.

   If the trajectory from Nowell to Speed could be described as the attempt
to slowly minimalize Irish difference, then a similar development is

recognizable in maps that do not take the entire island as their object but focus on specific regions – the single county, a strategically important valley, an isolated settlement. Such maps – showing confiscated lands, private estates or politically sensitive areas, and serving as legal documents, political aids or military tools – were of more immediate functional use than general descriptions of Ireland, and are thus more directly representative of the driving force behind contemporary cartographic production. Yet their essentially pragmatic functions cannot disguise that their purpose, in an Irish context, was nearly always also to celebrate English political supremacy in Ireland. Map-makers, that is, translated into the presentational code of the visual design those circumstances which enabled the material production of maps in the first place. In the remainder of this chapter I will discuss two examples of such maps: the military campaign maps of Richard Bartlett and the plantation survey of Thomas Raven. Both examples focus on Ulster, still predominantly Gaelic in the early seventeenth century and hence a physical and discursive terrain where cartography most urgently needed to demonstrate its superior vision.

As already noted above, Richard Bartlett was not to survive his sojourn in Ireland. In meeting his violent death at the hands of the inhabitants of Donegal, who evidently thought little of being written off their land by an English map-maker, his fate as a cartographer is not without a certain degree of historical irony: from the empty and depopulated space of his maps the native Irish reappeared in flesh and blood to emphatically bring their own perspective to bear on this paper version of Irish spatial realities. Bartlett originally came to Ireland as a member of Mountjoy's expedition which, in 1603, effectively marked the end of the Nine Years War. He thus participated in the final overthrow of an almost decade-long rebellion during which Hugh O'Neill had mounted such a significant military challenge to the English forces that '[f]or the first time since the original Anglo-Norman excursion into Ireland in 1167, England faced the prospect . . . of losing control over the entire island of Ireland.'[34] Most of Bartlett's maps were quite clearly intended as straightforward celebrations of Mountjoy's campaign and it is principally the story of this military success which his maps set out to relate in graphic detail.[35] My emphasis on 'story' is not accidental: what is so striking about Bartlett's military maps, and his colourful bird's eye views, is that they do not so much record the given – however ideologically constructed – situation of a particular moment in space and time as characteristically employ a narrative mode of representation.

The map of Moyry Pass (Plate 12) shows a regional map on the top half and concentrates in the bottom half on a particular detail of this landscape – Moyry Castle which had only recently been built by Mountjoy's troops to secure the passageway leading from the south into the Gaelic-dominated northern province of Ulster. While difficult to locate on the top half of the map Moyry Castle is shown in full prominence on its lower section where it majestically commands the surrounding countryside. Although built in only one month, and thus hardly as impressive a structure as the image implies, the selection of this particular detail of the landscape visually demonstrates English dominance over Irish soil: the cartographer suggests that the English have (or ought to have) full military control over this region. The pictorial arrangement guides the viewer's eyes: Moyry Castle, highlighted so markedly and with St George's flag flying over the tower, provides the visual perspective from which to survey the country; the English castle has become, in terms of the cartographic display, the dominant spatial paradigm of the landscape depicted. This map – like much of Bartlett's work a deliberate propaganda piece – constructs Ireland as a military landscape under tight English supervision, but the visual arrangement as a whole cannot fully conceal what necessitated both the map and its referent – the military stronghold – in the first place: the menacing presence of the invisible Irish, hiding behind the next drumlin or gathering their forces in the adjacent valley.

Bartlett's description of the Irish castle of Dungannon and the O'Neill inauguration stone at Tullaghoge (Plate 13), a map without a single toponym, takes the symbolism of the Moyry Pass map even further. This map, again, is a narrative of conquest. Dungannon castle was the central stronghold of the O'Neills but is now – as St George's flag indicates – safely under English control. While Bartlett places the conquered Irish castle in the centre of the map, signifying English victory, the military action that preceded this forceful appropriation is also represented, inviting the viewer to read the map as a sequence: Mountjoy's well-ordered camp is displayed in the bottom right-hand corner preparing the attack, the top inlay shows a scene of heavy military fighting with the native Irish defending their crannog (an ancient Irish lake-dwelling) against an overwhelming English attack. The stone chair at the very bottom of the map is the O'Neill inauguration stone at Tullaghoge, 'the territorial source and fountain of the kingship of the O'Neills'[36] – an ideologically highly charged location eventually destroyed by Mountjoy. The presence of this stone chair on the map – a Gaelic icon shown in all its prominence – may momentarily unsettle the cartographic script of English dominance. Yet ultimately overruled by the course of history, it is displaced to the right of the overlay surrounding the image which will eventually cover it once fully spread out: as a cartographic record of English triumph, the symbol of Gaelic supremacy disappears into insignificance.

Bartlett's maps are unusually suggestive for another reason. They all foreground the material quality of the cartographic image by deliberately highlighting the parchment on which the map is drawn. On top of the central image one or two additional maps, usually torn and scrolled at the edges, are superimposed – a representational mode which emphasizes the essentially textual character of cartographic images. These arrangements manipulate the cartographic sense of distance, bringing the viewer closer to selected landscape elements while holding others at bay. Establishing either a hierarchy of space – placing Moyry Castle in a central position in the landscape – or introducing a temporal sequence – detailing the stages of the campaign that led to O'Neill's defeat – this practice of spatial collage is as much a self-reflective experiment on the representational power of maps as an attempt to visually amplify English sovereignty over Irish soil. Further, since each map brings together various different locations in a single image it establishes spatial connections that had been invisible, in fact non-existent, before. The order of space, as represented on these maps, is not a neutral concept but the result of purposeful political intervention. On Bartlett's maps, land is rearranged to offer images of historical beginnings and to promote the contemporary political agenda: the cartographic image is the arena where the plot of the successful English conquest is set. Ireland ultimately emerges as an arbitrary textual collage of different maps.

In the early seventeenth century, after completion of the military conquest to which Bartlett's maps contributed, a demand for different kinds of maps arose. The central administration were obviously interested in obtaining geographical knowledge about areas hitherto beyond their immediate influence and many of the new English and Scottish settlers 'who acquired lands in Ireland during [that period], whether as part of an official plantation scheme or as ad hoc colonizers, needed information as to the extent and economic potential of their new lands.'[37] This situation turned Ireland into a profitable terrain for English surveyors. One representative of this new professional group, Thomas Raven, first appeared in Ireland in the early years of the Ulster plantation. In 1622 he was at work surveying and drawing maps of county Londonderry – so called because the entire county had been granted to the twelve

London livery companies. His maps were to be part of a more general survey of this area, a project initiated by Sir Thomas Phillips as an attempt to prove to the king that conditions of the original grant had not been observed by all planters.[38] Phillips' intentions explain why Raven, who had for years been employed as the city's surveyor, was initially unsure about his participation in the project. Writing to Phillips in 1621 he states that 'if the description [that is, the maps] be as you desire, it must needs lay open many things to his Majesty's view that may prove very prejudicial unto them [the planters]'. He therefore asked to 'be thereunto commanded [rather] than to be seen to take it in hand upon my own accord',[39] obviously thinking that the duty of office would relieve him from the stigma of disloyalty.

Despite this initial reluctance on the part of the surveyor, Raven's maps offer excellent insights into the specific mode of conceptualizing space in a plantation context. On these maps, 'unique for the period prior to Petty's *Down Survey*',[40] unrivalled prominence is given to the political and economic inscriptions on the landscape generated by the forceful change of ownership that had occurred prior to the settlers' arrival. At the centre of Raven's cartographic concerns are not the natural features of the landscape but the new proprietary boundaries imposed on the regional topography. This is, of course, quite generally the purpose of the estate map to which these plantation maps come closest in genre. As we have seen, estate maps were intended to show in detail the extent of a particular landowner's property and rarely expressed any interest in the surrounding landscape.[41] I noted in Chapter 2 that estate maps began to flourish in sixteenth-century England but transferred to an Irish context where land had only recently changed hands and the native Irish far outnumbered the settler community such maps take on additional meaning. Their general insistence on clearly defined boundaries ties in with the newly acquired confidence that had set in after the completion of the military conquest: on maps of earlier plantations woods and bogs featured prominently as possible hiding places of Irish 'rebels'. But in Raven's case the smaller the area depicted on the maps and the more the land represented is directly linked to individual settlers the more he loses interest in recording the distinctive regional topography. While the Sperrin Mountains still dominate the landscape on his outline map of the entire county (Plate 14),[42] the fanciful trees on the map depicting the vintners' lands (Plate 15) give the impression of being mere decorative embellishments. On his map of the 'Vintners' buildings at Balleaghe' (Plate 16) – showing no more than one settlement consisting of a row of houses, the fort (or 'bawn') and the church – the natural topography (apart from the river) has disappeared completely.

On this last map it is exclusively the man-made additions to the landscape that are of interest: each building seems to be rising out of a blank and homogeneous space, an untroubled expanse of virgin ground, while angles of vision, size of geometrical scale employed, and the perspectival relationships between buildings appear to be utterly arbitrary. Topographical concerns disappear behind the attempt to display the planters' material achievements and the configuration of their mental world: the precise and orderly arrangement of the buildings on the main street, parading the individual names of their owners, reflects the virtue of the planter community; the strong walls of the fort and the well-kept church indicate the settlers' strength and piety. What matters on these maps is not so much the nature of the surrounding countryside but shape and size of estate and buildings, marking the way the landscape has been successfully transformed into a new cultural space, emptied of all traces of an earlier history. The 'cosy' world of the planter community[43] on the map of Balleaghe is the result of a forceful act of erasure: nature and native population alike are written off the land, and the historical struggle that preceded this historical moment passes unmentioned. But even this map cannot completely silence oppositional voices. Hidden in the legend in the bottom left-hand corner are fragments of a reality at odds with Raven's well-arranged surface. The legend reads: 'British men present on these proportions: 80, whereof armed: 66. Natives on these proportions: 184.' It would be misleading to assume insurmountable antagonisms between planters and natives who frequently, much to the dislike of the English government, entered into contractual arrangements with each other.[44] But as a voice from the margins, disembodied and dispossessed, this verbal reminder of the presence of the otherwise invisible native Irish runs counter to the pictorial representation of the land which appears almost entirely purged, as it were, from any traces of the native population.

The topographical maps I have looked at in this chapter all demonstrate the desire to spatially appropriate Ireland and to absorb its landscape into the territorial concept of national British unity, yet each needs to register, to different degrees, the presence of older claims to the landscape. Identifiable as a conspicuous absence or a contained presence, as a deliberate pose of defeat or a textual margin, the native Irish cannot be fully erased from the surface of the map. But in each case the act of cartographic visualization reinvents the geographical notion 'Ireland' as a depopulated and empty, dispossessed and conquered space that could be properly mapped and subjected to new, politically charged, topographical visions.[45] Pre-colonial reality literally – as the case of Bartlett so vividly demonstrates – disappeared underneath the heavy layers of the colonizers' maps. John Donne's poetic description of a map-maker's workshop seems an apt analogy for this colonial logic behind the mapping of Ireland: 'On a round ball / A workman that hath copies by, can lay an Europe, Afric, and an Asia, / And quickly make that, which was nothing, all.'[46] Before the map (or the globe), there was indeed 'nothing': a white sheet of paper signifying the absence of culture rather than the existence of difference. Irish maps function precisely in this way: they do not describe, they create, they turn what is a *tabula rasa*, an empty slate, into fully colonized spaces. The map-maker 'quickly make[s] that, which was nothing, all' by giving material existence in the shape of a map to a particular conception of the land that aspires to stand as part of the national territory, the cartographic conquest signifying mental and political appropriation.

Part III
Narratives

Introduction

In his *Exercises* of 1595, a collection of didactic tracts addressed to 'all the young Gentlemen of this Realme',[1] Thomas Blundeville included a treatise on a large, 18-sheet world map by Petrus Plancius first printed three years previously in the Netherlands[2] and described as '[a] Mappe meete to adorne the house of any Gentleman or Marchant that delighteth in Geographie'.[3] The lengthy account, intended partly to promote technical understanding of modern cartographic projection, translates the visual landscape of the map into the verbal detail of a narrative description designed to recreate textually the earthly *varietas* of Plancius' image of the world. Based on a French original which accompanied the Dutch print, this vivid document of contemporary geographical thought embraces the ideal symmetry of the terraqueous globe in which the three continents of the old world – Europe, Africa, Asia – correspond textually and visually to the newly 'discovered' continents of Mexicana, Peruana and Magellanica. Pride of place is awarded to Europe which 'exceedeth all others in noblenesse, in magnificencie, in multitude of people, in might, puissance, and renowne', and culminates in the island of Britain, 'the greatest and mightiest of all *Europe*'.[4] The construction of Britain as the focal point of European geography serves to translate its insular existence into a sign of natural superiority – a textual move echoed hyperbolically in a sonnet by Robert Devereux, second earl of Essex, who defines the island as a modern *omphalos*, the pivotal point of an Anglocentric cosmography: '[s]eated between the Old World and the New',[5] Britain constitutes a sphere of its own, bound to neither side of an Atlantic world.

The ease with which the small scale of global representation could lead to the construction of new symbolic cosmographies demonstrates that the rapid expansion of geographical knowledge did not simply force contemporary spatial awareness into the framework of a 'scientifically' neutral world-view but released imaginative energies which, in early modern English writing, culminated in the fervent textual construction of a 'sceptred isle'.[6] Transferred from the periphery of the ancient world to its vanguard position on the edge of maritime Europe, facing new worlds across the ocean, the old Virgilian trope of geographic isolation – *toto divisos orbe Britannos*[7] – now served to shape the perception of Britain's insular uniqueness. As Lesley Cormack has recently claimed, Elizabethan geography encompassed both 'an expanding globe and an enclosing nation' and thus enabled Englishmen to 'increasingly [define] themselves and their country as separate from the Continent and the rest of the terraqueous globe.'[8] Robert Devereux' striking poetic image of an Atlantic Britain, celebrated in its insular separation as a 'land . . . no other land may touch',[9] thus not only evokes the oceanic space of the Ortelian world map but also provides a point of access into the logic of chorography and its desire to identify geographically the unique and special status of Britain. A mode of geographical thought concerned not with the global frame but with local detail, chorography might be the conceptual opposite of Plancius' cosmography, yet the need to focus on regional variation even in a comprehensive overview suggests the prevalence of discursive interrelations. The exchange between both geographical registers is also implied by the frequent reliance of the terminology of local description on the rhetoric of universality – exemplified, for instance, by John Norden's claim, in his *Description of Middlesex* (1593), that '[t]his our famous BRITANNIA' constitutes 'of it selfe another world'.[10]

But such mutual borrowings and conceptual overlaps demonstrate not only the interdependency of various levels of geographical thought, they also bring into focus the susceptibility of the cartographic image to its ideological appropriation by the logic of narrative. Blundeville, Essex and Norden all articulate their inflated sense of Britain's national uniqueness by forcing cartographic *visualizations* into *narrative* frameworks. Translated into textual causality, their map of Britain governs either the structural harmony of the earth's six continents, mediates between two gigantic land masses, or concentrates in a single island the spatial energy of the entire globe. Such imperial claims are recognizably generated by the visual power of specific maps: Blundeville's *Exercises* openly offers up Plancius' world map as its immediate referent; behind the 'omphalos syndrome'[11] of Essex's poem lurks a palimpsest of Ptolemaic and Ortelian cosmographies; and Norden's county description is suffused with the universalizing claims of Saxton's *Anglia*. One vital political effect of this narrative appropriation of the topographical map is immediately apparent, for the series of spatial translations from the geometric to the visual to the verbal, from number to image to text, reverses the script of national inferiority which featured strongly throughout the final years of Elizabeth's reign – still audible, for instance, in Thomas Hood's anxious evocation of the 'proud disdayn-full insolent Spaniard':[12] the topos of insular separation, the image of peripheral Britain as 'a world apart', no longer defined a secluded island kingdom spatially removed from the centre of historical action, but forcefully affirmed British imperial aspirations as a powerful and independent political agent in a global, transcontinental scenario. John Donne, echoing Essex, famously applauded this shift when he thanked the Virginian planters for '[having] made this *Iland*, which is but the *Suburbs* of the old world, a Bridge, a Gallery to the new'.[13]

If the defining property of narrative is its inherent temporality, then the explicit intersection of time and space in 'literary' maps is hardly surprising. 'Welcome destruction, blood, and massacre! / I see, *as in a map*, the end of all',[14] cries Elizabeth in *Richard III*, anticipating in a prophetic moment the havoc wreaked on England by Gloucester's greed for power. Serving as an index to future events, Elizabeth's apocalyptic map shows coming bloodshed, not current topographies. Hers is a map of time, not of space. In Parts I and II, I have noted at several points how topographical maps may equally assume such temporal qualities in their interaction with the mental and material conditions of their historical moment, how the cartographic picture plane, to put the matter more succinctly, may inscribe territorial with narrative 'plots', if by virtue of its semantic context or its visual design. The Ortelian world map, for instance, is deeply implicated in a colonial narrative of erotic/exotic desire, while a structurally similar process of narrative appropriation resulted both in Speed's cartographic investment of national space with a specific political agency, and in Bartlett's inscription of colonial Ulster with tales of English conquest. If these examples show that the visual and textual properties of maps cannot be easily divided into two distinct groups, they also show that the iconic quality of topographical imagery always points beyond itself, always produces additional meaning by creating narrative causalities which surpass any map's immediate physical reference. The present section, Part III, addresses more fully the questions arising from this dialogic fusion of the visual and the verbal, from the subsumption of the spatial image under the temporal sequence of narrative, through an examination of the spatializing strategies of two literary modes of 'mapping', chorographical writing and epic poetry.

The chapter that immediately follows looks at chorographical narratives focused on Britain. As a genre of geographical writing chorography owed its imaginative perambulation of the nation to the semiotics of the map which in turn relied on the topographical and antiquarian information supplied by the textual account. But the written description of Britain could take different forms. The chorographical literature that precedes the publication of William Camden's *Britannia* (1586) – most notably William Harrison's *Historicall Description of the Islande of Britayne* (1577) – offers an imaginary national geography that is not based on the influential political choice of the county as the relevant representational unit of chorographical description. Elaborating on the conceptual distinction between Camden and Harrison's topographical projects, Chapter 8 then contrasts Edmund Spenser's national epic *The Faerie Queene* with Michael Drayton's chorographical poem *Poly-Olbion*, two vast poetic versions of national space embracing rival theories of the cultural function of cartographic representation. The final chapter continues the juxtaposition of national and colonial space by examining the narrative inscription of Ireland with discourses of barbarism. This discursive effort, generating a range of functional tracts promoting Irish political reform, was instrumental in producing Ireland as a liminal cultural space, continuously meandering between its signification as a barbaric outpost and a territorial extension of the emerging nation.

7
Imaginary Journeys: Describing Britain

The term chorography, in the basic definition deriving from Ptolemy, applies to the description of particular regions, defined against accounts dealing with whole continents or the entire earth, which properly belong to the respective disciplines of geography and cosmography. Drawing on Ptolemy, William Cuningham explained that

> lyke as Cosmographie describeth the worlde, Geographie th'earth: in lyke sorte Chorographie, sheweth the partes of th'earth, diuided in them selues. And seuerally describeth, the portes, Riuers, Hauens, Fluddes, Hilles, Mountaynes, Cities, Villages, Buildinges, Fortresses, Walles, yea and euery particuler thing, in that parte conteined.

In an analogy that was equally thought to derive from Ptolemy, Cuningham suggested that the chorographer is best compared to a painter who 'shuld set forth the eye, or eare of a man, and not the whole body, so that Chorographie consisteth rather in describyng the *qualitie* and figure, then the bignes, and *quantitie* of any thinge.'[1] Cuningham's insistence on this distinction points to a wider conceptual difference between the various genres of geographical writing. Renaissance geographers, as Frank Lestringant has recently noted, increasingly came to realize that 'the small scale of global representation was radically distinct from the medium or large scale appropriate to a region, be it more or less extended, of the earth. The former grasped the *quantity* of the world, whereas the latter plumbed its *quality*.'[2] Different spaces thus required different frames of representation; and where the chorographer would measure the particular detail, the regional topography and the isolated event on the qualitative (and technically large) scale of local description, the cosmographer would operate within a vast global

137

scenario, charting the size, shape, and quantity of continents on a scale so small that the brooks and forests of a national landscape never even came into view. Chorography, that is, dealt with lived spaces, cosmography with the purely conceptualized, literally *invisible* realms of the imagination.[3]

With hindsight, chorographical writing may appear as little more than a dull exercise in cartographic *ekphrasis*. To be sure, the dryness of much of the extant material hardly conveys the sense of excitement that accompanied this innovative interest in Britain's national topography. In 1950, A. L. Rowse drew an explicit verbal parallel to the impact of the New World on European spatial consciousness by identifying the English chorographical project as 'the Elizabethan *discovery* of England'.[4] Describing the motivation behind what he saw as a concerted effort of 'chroniclers and topographers, ... antiquarians, surveyors and map-makers', Rowse wrote in glowing terms about the 'increasing self-consciousness of the nation', the 'spirit of national independence' and the country's emergent 'sense of itself'.[5] These passages from Rowse's currently rather unfashionable study of the late Tudor period, *The England of Elizabeth*, are perhaps more remarkable for the author's own patriotic investments in a glorified national past, than for their observations on the Elizabethan conception of nationhood, but serve as a useful reminder that the most convincing contemporary documents of national self-representation emerged from the practice of geography, not history. If England – or rather Britain – was the object of a 'discovery' in the sixteenth century this was achieved less by narrating its progress through time than by describing its expansion in space: on Rowse's list only the chronicler operates within a temporal rather than a spatial framework.

And yet, written chorographies, especially those focusing on counties, tend to be little more than flat descriptions of the countryside, long exhaustive topographical accounts – or merely inventories – which need many pages of writing to rival what maps appear to achieve effortlessly in a single image. Arguably, though, this comparison between texts and maps does not do justice to the difference between the respective persuasive powers of cartographic image and spatial narrative. In its reflection of a world 'out there', a map needs to import into its visual display certain properties of the original it purports to represent;[6] indeed, its claim to convey an 'accurate' representation of geographical fact rests on the degree of identity it manages to establish between itself and its object of description. Narrative accounts of space, however, rely on rhetoric, on the implicit understanding that what is classified as the 'real' is nothing but the raw material of a linguistic effort that aims to reproduce the substance of the world as a string of verbal figures. Within their preconceived representational forms written chorographies could thus shape images of the physical world with more interpretative freedom even than maps. The difference between the two textual models most frequently employed by topographical narratives corresponds to the conceptual distinction between plan and itinerary – two descriptive modes informed, respectively, by the external, static knowledge of the chart and the internal, dynamic experience of the voyage. Neither textual form supplied simply a neutral geographical frame. In the description of Britain, opting either for systematic control (of the plan) or operational uncertainty (of the journey) was never merely an aesthetic choice. In each case the resulting narrative entailed a moral and political outlook which shaped a specific version of national space at the cost of excluding another. And as we shall see, in descriptions of Ireland, where the ground level experience of Irish difference yields to its rhetorical effacement from above, the spatializing strategies of itinerary and plan are again enlisted, though in rather unexpected ways, to support English political reform plans.

What, from a contemporary perspective, was the use value of a map? One of the most extensive answers to this question was offered by Cyprian Lucar whose manual on surveying is ample evidence that the analysis of maps should not restrict itself to a discussion of their topographical data. For Lucar a map needed first to inform about the local practice of agriculture, the situation of the ground, vegetation, forests and parks, then about harbours, ports and other landing-places, the size of the king's navy kept there, the tidal rhythm 'and whatsoeuer else that said described land hath strange, new, notable, and commodious.' Maps should also include information about houses and buildings, settlements and fortifications, climatic conditions, rivers, bridges and roads, the legal status of towns, 'whether the people there are wittie and of quicke conceite', their moral virtues, the number of inhabitants and soldiers, the extent and storage of their weaponry. Lucar ends by urging the cartographer to mention 'whether the said described place is by nature, or art so scituated, as that it cannot be scaled with ladders, beaten downe with great ordinance, or vndermined to be blowne vp with gunpouder'.[7] Evidently, maps also helped destroy what had taken such effort to describe in detail. The excessive demands made on the map-maker's skills in this passage illustrate that the process of transferring world into map required more than proficiency in technical surveying. Lucar does not conceive of maps as geometrical accounts of superficial topographies but as strategic instruments of knowledge whose function was to graft an economic, political and ethnographic script onto their topographical information. Yet his comments clearly exceed the concept of a map as a visual image and their relevance is more readily grasped when applied to textual rather than cartographic descriptions: it is precisely the written chorography, a genre that began to flourish in late Tudor England, that contains much of the type of information Lucar claimed to belong to the domain of the map.

As a genre, such topographically organized accounts of individual regions were defined against the political narrative; the chorographer's method was antiquarian, not historical. William Lambarde, in his *Perambulation of Kent*, stated that it was his 'purpose speciallye . . . to write a *Topographie*, or description of places, and no *Chronographie*, or storie of times'.[8] Concerned with space rather than time, with place rather than person, late Tudor and early Stuart antiquarianism looked back to the pioneering work of John Leland, royal librarian under Henry VIII, who had assembled, on a nationwide itinerary lasting several years, a vast collection of notes intended to form the basis of a multi-volume chorographical description of Britain. In a promotional tract presented to the king (and later edited by John Bale) he described himself as 'totallye enflamed wyth a loue, to se throughlye all those partes of thys your opulent and ample realme'. His subsequent travels were so exhaustive, he wrote, that

> there is almost neyther cape nor baye, hauen, creke or pere, ryuer or confluence of ryuers, breches, washes, lakes, meres, fenny waters, mountaynes, valleys, mores, hethes, forestes, woodes, cyties, burges, castels, pryncypall manor places, monasteryes, and colleges, but I haue seane them, and noted in so doynge a whole worlde of thynges verye memorable.[9]

Leland's project, unprecedented in scope and intent, clearly centred on the land itself as the source for a description capable of capturing the essence of Britain. But confronted with an undigested mass of data Leland was unable to give his discoveries a coherent textual structure. His fate, contemporaries reported, was to go insane over the vast material he had collected.[10] Despite his failure Leland inspired much subsequent chorographical writing and was acknowledged as a central source in both Harrison's *Description* (1577) and Camden's *Britannia* (1586), the two most substantial national chorographies to be written in the latter half of the sixteenth century. Leland's description was intended to embrace a comprehensive 'British' view of its subject matter, a model adopted (and even expanded to include Ireland) both in the *Britannia* and in *Holinshed's Chronicles* (which opened with Harrison's *Description*). Yet although Harrison and Camden shared Leland's emotional response to his native soil and continued, in a sense, his search for an appropriate textual format, their respective constructions of the social space of the nation operate within opposing conceptual frameworks.

Following the medieval model of prefacing historical narratives with topographical introductions,[11] Harrison's *Description* offers 'a broadly imagined, geographical, social, and institutional account of a country, whose function is to lay down the base for the history to follow.'[12] The text provides a spatial setting for the ensuing chronicle but moves beyond a mere illustration of England's topography[13] by addressing a variety of social, economic and political themes, ranging from topics of national interest – the structure of society, the legal system, the distribution of fairs and markets, and so on – to regional peculiarities, like the 'infinite number of swannes'[14] near London Bridge or the Halifax public execution ritual which requires the active participation of the local populace. Though dominated by a patriotic stance the text is frequently interspersed with complaints about England's inner decay and the moral corruption brought about by cultural and economic change. The material, Harrison explained, was gathered from a variety of sources. He had consulted all available reference works, among them Leland's notes ('bookes vtterly mangled, defaced with wet, and weather'[15]), corresponded with residents throughout the country and discussed his findings with his fellow antiquaries. The result of this scholarly exercise is a topographical description that attempts to marshal its material through a loose sequence of thematic sections: 'of cities and towns', 'of palaces belonging to the prince', 'of the building and furniture of our houses', 'of prouision made for the poore', are typical chapter headings. In moving across a multi-faceted spatial reality, the *Description* looks at topics as diverse as England's architecture, its inns and fairs, forests and parks, law courts and universities; it describes local customs and the dietary habits of its population; it includes passing observations on daily life and bitter complaints about the vagaries of contemporary fashion; it sketches out the history of England's political landscape, its ancient road system and current clerical and administrative divisions.

Topographically, the text is organized around an account of England's rivers, a dominant descriptive convention in the chorographical tradition. Rivers are the dynamic element of landscape, they produce the movement and fluidity chorography requires in order to overcome the impression of representational stasis. When Harrison, an armchair geographer who never ventured far beyond his local parish, describes his own work as an imaginary journey ('I sayled about my country within the compasse of my study'[16]), the travelling metaphor is not mere poetic flourish but a constitutive principle of the dynamic space constructed in the text of the *Description*. Bringing the first stage of his textual river journey to an end, Harrison looks back over the terrain his writing has covered: '[F]rom the hauen of Southampton, by south vnto the Twede, that parteth England and Scotland, by north (if you go backward contrary to the course of my description) you shall finde it so exacte, as beside a fewe bye ryuers to be touched hereafter, you shall not neede to vse any further aduise for the finding and falles of the aforesayd streames.'[17] Sending his readers backwards through the linear sequence of his chapters, Harrison gestures at the multiplicity of possible journeys across the map of the nation, through its physical, cultural and political space. But although the second edition of *Holinshed's Chronicles* includes a direct reference to Saxton's atlas at this point, Harrison's textual cartography cannot be fully reduced to the figure of the map. In tracing England's rivers across the national territory it is interlaced with another mode of representing space, the itinerary. The description of place, according to Michel de Certeau, 'oscillates between the terms of an alternative: either *seeing* (the knowledge of an order of places) or *going* (spatializing actions).'[18] The former is the privilege of the map, projecting on to a plane, and presenting to full view, a totality of spatial relations; the latter belongs to the order of the itinerary or tour, exploring space through movement and operative action. Harrison's *Description* fluctuates between these two moments, between a space planned and centrally regulated, and a space of lived experience, shaped and used by its inhabitants, defined principally through its social dimension.

Against this background of a structural network of routes and trajectories across England's topography, Harrison's 'poeticall voiage'[19] takes the reader through a seemingly incoherent series of thematic chapters, foregrounding a degree of diversity and fluidity that challenges any monolithic view of the nation's cultural and social – and ultimately spatial – configuration. The text provides the elements for alternative narratives, other discursive voyages, that might convey the experience of a complex social realm at odds with the dominant royal perspective inscribed in the massive chronicle history that follows. In offering cultural diversity as a structural principle of its textual progress, the *Descrip-*

tion is perhaps better considered a conceptual alternative to, rather than merely the imperfect precursor of, William Camden's hugely successful *Britannia* whose first publication in 1586 coincided to a year with the second (and last) contemporary edition of *Holinshed's Chronicles*. The *Britannia* grew vastly in bulk over the years and the substantially revised Latin edition of 1607, translated three years later by Philemon Holland, 'stands as the central achievement of the English chorographical tradition.'[20] Camden's conception of national space corresponds to the synthesis of a unifying cartographic order. Like Saxton's atlas, the *Britannia* seizes on the county as the central unit of chorographical description, devoting a full chapter to each: the partial view of the individual shire validates the overarching national frame of reference. The horizontal division of a political landscape, essentially an administrative pattern, thus precedes all further detail on British topography. Masking its textual linearity, its successive unfolding in time, the *Britannia* treats space from the outset as the object of an immediate knowledge, imitating the totalizing approach of the map: as if at every stage of the narrative the space of the nation is 'always already visually present, fully offered to full view and potential speech.'[21]

In opening the voluminous tome with a historical section that introduces the much longer perambulation of the individual counties, Camden turns on its head the medieval model *Holinshed's Chronicles* had adopted: in the *Britannia*, history sets the stage for geography. It may indeed be 'difficult to give an adequate indication of the contents of the *Britannia*',[22] yet the topographical descriptions of the counties follow a more or less coherent format. Outlining his agenda, Camden notes that

> [i]n the several discourses of every of [the Provinces or Shires of Britaine], I wil declare as plainly and as briefly as I can, who were their ancient Inhabitants; what is the reason of their names; how they are bounded; what is the nature of the soile; what places of antiquity and good account are therein; what Dukes likewise or Earles have beene in ech one since the Norman Conquest.[23]

Preoccupied with names and boundaries, with genealogies and the continuity of human settlement, the *Britannia* streamlines its topographical and antiquarian information into the celebration of a landscape shaped by successive generations of the leading families of the gentry. Even when the description follows a county's rivers, these are shown to be flowing exclusively around stately mansions, ancient castles and private parks. The equation of national space with the realm of a social elite, of which a historicized landscape bears witness, guarantees the stability of a political order and allows its translation into the static coexistence of individual plots on the imaginary plane surface of cartographic projection. In contrast to the mobile landscape of Harrison's *Description*, space in the *Britannia* is less the product of social interaction, defined across a wide spectrum of possible readings, than a political narrative claiming the nation as the exclusive domain of its landowning 'Dukes' and 'Earles'.

When derivatives of Saxton's maps were included in the *Britannia*, from the 1607 edition onwards, Camden implicitly acknowledged the need to subject the visual image of space to the political perspective laid down in the textual descriptions. 'Many haue found a defect in this worke', he explained in the preface, 'that Mappes were not adioined, which doe allure the eies by pleasant portraiture, and are the best directions in Geographicall studies, especially when the light of learning is adioined to the speechlesse delineations.'[24] Maps visually offer up landscape for discursive appropriation: they are pleasant, but speechless; they allure the eye, but are without the 'light of learning'. If the nation is defined in and through space, then its cartographic representation needs to concur with the precepts of the dominant political narratives surrounding its production. As we have seen, *The Theatre of the Empire of Great Britain* proceeded along similar lines and though Speed's emphasis was on the map, rather than the text, his set-up echoes the post-1607 editions of the *Britannia*. Such textual and pictorial strategies focus the reader's gaze squarely on the land and its history, 'direct[ing] attention away from the king and toward the country',[25] and thus helped, despite the overt congruence between Speed's programme and James' political agenda, to challenge a purely dynastic identification of nationhood. But in employing the land itself as the central element of historical stability, these gentry-oriented representations describe Britain as a static spatial (and hence social) order, markedly at odds with Harrison's vibrant cultural landscapes. Camden and Speed enlist geographical discourse in order to construct a national topography as the correlate of a social space compatible with the political interests not of a broadly defined national community, but of a narrow 'brotherhood' of landowners.

The subsequent development of chorography saw a fragmentation of the national vision that characterized the work of Harrison and Camden, Saxton and Speed. Chorographical writing increasingly focused on individual counties, marking a clear shift 'from the kind of antiquarian study that represented a common, national heritage to a much more exclusive focus on individual ownership of the land'.[26] The resulting descriptions – the various Jacobean and Caroline surveys of Devon or Suffolk, Leicestershire or Staffordshire – tended to be little more than sterile inventories that would collect genealogical information and establish an order of property that firmly linked the land it described to its resident gentry.[27] Such static archival exercises owed little to the imaginative conception of space that marked the early phase of the chorographical project. But they hardly constitute a surprising sequel either, for the descriptive activity of chorography had always implied a relationship to the land that relied more on preconceived representational patterns than on an immediate personal acquaintance with the national territory. Leland may have visited every single parish in the land but his project failed precisely because he was unable to develop a systematic textual frame for the vast quantity of data he had collected. Saxton's view of the English and Welsh countryside was exclusively organized by his theodolite, or whatever surveying instrument he may have been using, and the main reason Camden travelled so extensively across Britain was to consult further source material. Speed never bothered to journey into all the corners of his maps[28] and Harrison had no qualms about openly confessing to his patron that 'except it were from the parish where I dwell, vnto your Honour in Kent, or out of London where I was borne, vnto Oxforde and Cambridge where I haue beene brought vp, I neuer trauailed 40.miles in all my lyfe'.[29] Evidently, for the sixteenth century, the geography of the nation still served more as a discursive than a physical terrain in which to conduct the search for its political and cultural meaning. If the emergent nation appears as an unstable signifier in contemporary maps and texts, if neither its outward form nor inward structure find their definite shape, these uncertainties only serve to demonstrate that there was more than one Britain, more than one conception of national space, to be imagined or 'discovered' in the productive arena of early modern geographical discourse.

The extent to which these uncertainties could even prove a disabling factor in cartographic production may be briefly demonstrated by a look at a writer, chorographer and map-maker whose work occupies a central position in the history of English cartography, John Norden. When Norden wrote *The Surveyor's Dialogue* in 1607 he was already looking back on a varied career: he had started out as an estate surveyor, was a prolific writer of devotional tracts, and later moved on to produce a panoramic view of London. In the early 1590s he embarked on a chorographical project that was designed to combine visual and textual material relating to Britain's topography in a series of single county descriptions gathered under the general heading *Speculum Britanniae*. This project never properly got off the ground. Only two county descriptions appeared in print during Norden's lifetime, while some four or five others remained in manuscript. In 1604 the project was effectively abandoned and Norden turned again to estate surveying. Given the general popularity of English county maps, why did the *Speculum* fail? Certainly not for lack of technical skill. Norden was working from fresh surveys and the maps he produced were considerable improvements upon Saxton's efforts; the inclusion of roads and an elementary grid system for speedy reference were wholly innovative features. Speed, wherever possible, copied Norden's rather than Saxton's maps.

But despite the quality of his work Norden failed to secure the financial support that would have enabled him to continue his project. As Helgerson has recently noted,[30] the traditional explanations for this lack of patronage – that Norden's religious fervour was unwelcome to Burghley (to whom Norden frequently applied for money), or that he may have continued to support Essex after 1601 – explain neither why his pass had not been renewed long before Essex's doomed rebellion nor why his religious activities had not bothered the crown when Norden's project initially found Burghley's favour. Instead, Helgerson argues, Norden's project was caught up in an ideological tension which had emerged since the publication of Saxton's atlas and which had significantly transformed the political function of the map of the nation. This ideological tension was the result of the growing contradiction between the conception of a land-based national identity inherent in national chorographies, to which the *Speculum* seemingly belonged, and the pro-monarchical concept of the estate survey which acted as an emblem of dynastic hierarchy and thus reconfirmed traditional patterns of monarchical sovereignty. Turning down Norden's request for financial support was thus a deliberate decision against a specific project rather than a general expression of personal displeasure, a fact made more obvious by Norden's subsequent appointment as surveyor of the duchy of Cornwall.

But if the national map had become so fraught with ideological tension that the government actively sought to prevent further cartographic work wherever it could, the intervention of maps in a tense social sphere had also turned them into such contested spatial signs that the incompletion of the *Speculum* points to the very difficulties Norden must have experienced in controlling their representational force. At least this is what his own views in defence of the project suggest. In 1596 he published a short treatise entitled *Nordens Preparatiue to his Speculum Britanniae*, intended, as the subtitle explains, as 'a reconciliation of sundrie propositions by diuers persons tendred, concerning the same [the *Speculum*]'.[31] After rehearsing several technical aspects of the representational practice he had adopted in the *Description of Middlesex* (1593), the only one of his county descriptions then known to a general public, he moves on to consider what is in effect the social and political subtext of his topographical imagery. Discussing the question whether he had included too much or too little detail on his maps he argues that '[t]he more things (as I take it) are obserued, the more like is the discription to the thing discribed.'[32] This may seem like an appeal to common cartographic sense, but the semantic conflicts endemic in proposing that the sheer quantity of data will establish a mimetic relationship between map and landscape are not so easily reconciled. Competing views as to whether 'houses & other thinges of small moment' should be excluded in favour of 'some of greater worth to be remembred',[33] for instance, reveal the surface of the map as deeply implicated in a controversy over social authority: whose memory does the image address? Nor is the cultural and historical signification of the national landscape any easier to determine, as is evident from Norden's own discussion of English place-names. Wrong usage, false resemblance between words and imperfect memory have clouded in mist the ancient toponymic surface of Britain: 'The affinitie of sundrie determinations of the names of places, and the ignorance of their significations may (besides the vulgar vnskilfull instruction,) miscarrie men from the truth.'[34] These representational uncertainties result in imperfect, even untrue, maps, or maps that show a different nation from the one imagined by specific readers.

The nature of these conceptual difficulties that stalled Norden's project is finally suggested by a minor linguistic readjustment observable in the programmatic vocabulary that accompanied the *Speculum* over the years. In *Middlesex*, the first part of the prospective collection, Norden characterized his 'intended labours' as 'the *description* of famous ENGLAND'.[35] Three years later, when he defended his project in the *Preparatiue*, the overall title had changed to 'the *rediscription* of England'.[36] This almost insignificant addition of the two-letter prefix, I want to suggest, exposes a further representational conflict inherent in the project. The act of 're-describing' – that is, re-mapping and re-writing – the physical territory of England may not have generated any moral doubts about its national grandeur – as the figure of paradiastole, or 'rhetorical *redescription*', clearly did in the evaluation of moral action (as Quentin Skinner argues[37]) – but could produce disturbing uncertainties concerning the uniqueness of its social and political, as well as historical, meaning. The discursive instabilities invited by each single toponym, for instance, point to semiotic struggles Norden could not fully control – does the tag 'Waldt or Weldt' derive from 'a wilde Horse, a wilde Bore, *or a wild man*' formerly living 'sauagely'[38] in that place? Each etymological option, while occupying a spot in the same semantic field, contains a radically different history of landscape and settlement. On Norden's county maps, increasing social mobility constantly produced such semiotic uncertainties and left the cartographic surface riddled with ambiguity. As 'redescriptions', they unsuccessfully attempted to bridge the discursive gap that had emerged between rival claims to the 'ownership' of the nation, and their failure to reach print and secure the approval of a map-reading public epitomizes the location of the national map at the nexus of conflicting discourses which had turned cartographic images into contentious cultural statements.

8
The Poetics of National Space

In 1625, Norden finally achieved what had seemed an impossible project before, to have '[t]he whole Kingdome yet described by one and the same Scale.'[1] Having abandoned the *Speculum* two decades earlier, he did not accomplish this geometric synthesis in maps. The comment refers to another innovative product emerging from his workshop: a guidebook designed to facilitate travel within England. On the 36 tables that make up *An Intended Guyde for English Travailers* the distances in miles between England's major towns are listed systematically in the form of a triangular reference chart now common to every modern road atlas. The realization of this practical tool which reduced England in a few pages to a set of mathematical figures had not been an easy task, for

> it is not possible for any Artist, so precisely to deliniate so great (nay farre lesse) Countrey, and the perticuler Townes, and their seuerall distances within the same; but that some errours of necessitie will be committed, especially by reason of hills, dales, woods, and other impediments, which intercept the view from station to station. So that the lines of opposition cannot be so exactly directed, as vpon a plaine and open horizon. But were the distances neuer so truly taken, by the intersection of right lines, yet in riding or going, they may seeme vncertaine, by reason of the curuing crookednes, and other difficulties of the wayes.[2]

This passage registers an internal conflict between the irregularities of landscape 'intercepting' the suveyor's view – the 'hills, dales, woods, and other impediments' – and the unrestricted expanse of mathematical space – manifest in the 'plaine and open horizon' – a conflict that

suggests the distinction between the traveller on ground level, negoti-
ating the 'curuing crookednes, and other difficulties of the wayes', and
the privileged map user, surveying the entire country in a single view.
In setting up this contrast, Norden not only reminds us that travelling
and map viewing were two quite different things, he also – as I will
argue in this chapter – provides us with a blueprint for a conceptual
distinction between rival versions of the national landscape.

For if the topographical map inspired the abstract mathematization
of space in the *Intended Guyde*, it was also absorbed into larger narrative
accounts – poetic and otherwise – of the nation, and it is in the trajectory that leads from Spenser's *Faerie Queene* (1590/96) to Drayton's *Poly-
Olbion* (1612/22) that the wider cultural significance of the difference
between the various spatial constructs implicit in the national project
emerges. The distinction between the two epics[3] on which I will elaborate is partly analogous to the shift from itinerary to map which characterized the reworking of Harrison's fluid national topography into
Camden's rigid atlas structure. Yet by giving rise to a vision that
processes national space through mythological frameworks, the poetic
attention geographic material attracts in the work of Spenser and
Drayton pushes this duality to a higher level. Put briefly, I will argue
that *The Faerie Queene* persistently invites the reader to engage actively
in the dynamic performance of space, while *Poly-Olbion* unfolds its
national scenario against the secure background of a fixed and intransigent geographical order. The link between Spenser and Drayton is
most often described as that between master and student: if Spenser was
the Elizabethan prince of poets, Drayton was his most industrious dis-
ciple in the Jacobean age, the foremost 'Spenserian' poet of his time.[4]
Yet other than the nostalgic longing for a lost Elizabethan sensibility
governing *Poly-Olbion*'s productive context these two monumental
works show few signs of internal contact, and the present chapter looks
to explore not their similarities but the opposition between them in
terms of their contrasting conceptions of space. In light of the generic
affinity between Drayton's *magnum opus* and the chorographical tradition considered above, I will disregard chronological sequence and
begin my discussion with *Poly-Olbion*.

Like John Norden, Michael Drayton is famous for his complaints. He
did not suffer from lack of patronage, or feel the need to rail against
local residents who failed to supply him with the topographical information he required – accusations Norden rarely spared his readers in
letters and prefaces – but had a similar penchant for addressing his audience in terms of scornful reproach. The letter 'to the generall reader'

that prefaces the first 18 songs of *Poly-Olbion* opens with a reflection on poetry that has become almost axiomatic in distinguishing between Drayton's public conception of literature and the 'private', coterie verse of the metaphysicals: 'In publishing this Essay of my Poeme, there is great disadvantage against me; that it commeth out at this time, when Verses are wholly deduc't to Chambers, and nothing esteem'd in this lunatique Age, but what is kept in Cabinets, and must only passe by Transcription'.[5] Drayton clearly felt swamped by contemporaries who 'had rather read the fantasies of forraine inventions, then to see the Rarities and Historie of their owne Country delivered by a true native Muse', and treated them with little respect: *Poly-Olbion* was not for those 'possest with stupidity and dulnesse' who chose 'to remaine in the thicke fogges and mists of ignorance'.[6] The latent bitterness of the 1612 preface was amplified ten years later when Drayton had finally found a publisher for the 12 final songs of *Poly-Olbion* (after a frustrating search lasting four years). Sourly addressing it '[t]o any that will read it' Drayton reported that he had met only with 'barbarous Ignorance and base Detraction', and prophesied that 'such a cloud hath the Devill drawne over the Worlds Judgment, whose opinion is in few yeares fallen so farre below all Ballatry, that the Lethargy is incurable'.[7]

The tone of these two prefaces has significantly influenced the critical reception of *Poly-Olbion* up to the twentieth century. Variously read as accurate observations on contemporary taste which no longer appreciated the 'strange Herculean toyle' (30, 342) of the national epic, as the angry comment of an ageing poet out of touch with his time, or as the general expression of a Jacobean nostalgia for a glorified Elizabethan past, Drayton is still chiefly remembered for his 'belatedness'. The awkwardness and forced conceit of *Poly-Olbion* have certainly baffled many readers, and some critics have even suggested that its overpowering length – 15 000 mighty alexandrines – should not be mistaken for an indication of its representative status in the Drayton canon. Jean Brink, in the most recent monograph on the poet's work, thinks that 'we have paid too much attention to [Drayton's] chorographical poem and too little to his satire. *Poly-Olbion* . . . has always appealed only to the "curious antiquaries"'.[8] Whatever the reasons for its failure to attract a wider readership, it will be argued here that *Poly-Olbion* should not be dismissed so easily as the inexplicable slip of an otherwise decent enough poet. Its textual conception of national space is hardly as belated or backward-looking as the poem's popularity among 'curious antiquaries' seems to imply; and in its attempted fusion (yet actual separation) of various discourses focused on an emergent sense of national identity – poetic, historiographical, cartographic – it creates in the figure of the invisible 'Muse' a fitting emblem for the disembodied nature of modern functional space.

Next to its belatedness, the second well known 'fact' about *Poly-Olbion* is its apparent unreadability. Ironically, perhaps, for a geographical epic, the main reason for the difficulty the work presents to twentieth-century readers seems to be the complete absence of a 'plot'. *Poly-Olbion* is a descriptive account of the geographical setting of England and Wales which contains hardly a single character (with the possible exception of Drayton's 'industrious Muse', a kind of early modern tour guide accompanying poet and reader throughout the 30 songs). It is true that Drayton makes ample use of prosopopoeia and that the myriad of nymphs, dryads and naiads that inhabit the world of the poem turn into a 'cacophony of voices' where 'seas, rivers, plains and mountains proclaim their own worth, loves and rivalries',[9] but this curious landscape choir hardly ever exceeds its oratorial function to engage in a recognizable course of action. In writing *Poly-Olbion*, Drayton's object was at least partially a versification of Camden's *Britannia*, or rather, its 'digestion in a poem' – a project fully summarized on the title-page: *Poly-Olbion, or A Chorographicall Description of Tracts, Riuers, Mountaines, Forests, and other Parts of this renowned Isle of Great Britaine, With inter-mixture of the Most Remarquable Stories, Antiquities, Wonders, Rarityes, Pleasures, and Commodities of the same: Digested in a Poem*. The implicit equation in the word 'digested' of a bodily function with the mental accommodation of disparate data into a single text is hardly accidental. The governing trope of the poem, as various critics have noted, is the familiar Renaissance conception of *discordia concors* – the yoking together of opposites to produce structural harmony – which Drayton achieves principally through his reliance on somatic metaphors: in *Poly-Olbion*, the 'organizing form or conceptual myth' which 'define[s] the relationship of disparate parts to the whole even as they suggest its multivalent meanings' is the image of 'the human body, a peculiarly apt metaphor for the kind of organic unity Drayton wishes to predicate of England.'[10] Describing the land in terms of the human body, the *chorography* of England is subtly *choreographed*[11] to figure as an animated landscape, producing – as on the book's frontispiece[12] – a composite image which weaves isolated natural features at every turn into the broad canvas of a unified whole.

The title-page announces two further structural features of the poem, the precedence accorded to geography (not history) in the national imagination, and the use of praise as a principal rhetorical strategy. As

Drayton's Muse proceeds on her journey through England and Wales (Scotland was to be the topic of a third instalment which remained unwritten[13]), it is the immediacy of its physical landscape, the *'Tracts, Riuers, Mountaines, Forests'*, that inspires the recollection of what is *'Most Remarquable'* about the past inscribed on its surface. Place consistently activates historical memory, a poetic design that results in an inseparable fusion of space and time: England and Wales turn into poem through an evocation of their geographical territory but the chronology of history needs to be imposed on the land, a mutually reinforcing link that spawns a national itinerary considerably different from the systematic perambulations of Camden or Speed. The first section of the poem, for instance, opens with the mythical landing of Brute in Cornwall. This is where the *Britannia* had also started, but unlike Camden, who in his opening book went on to cover all English shires in one systematic linear movement, Drayton's Muse first travels through Wales to teach us about Albion's 'British' past while later songs pursue an irregular and informal itinerary through the remaining counties of England, continuing the historical narrative up to the Elizabethan period and commenting in turn on the invasions of the Saxons, the Danes and the Normans. This parallel exploration of temporal and spatial trajectories through the 'renowned Isle of Great Britaine' converges in a ceaseless celebration of a rich and varied countryside. Diverse landscape features – hills and valleys, rivers and forests, bays, pastures and plains – all get the chance to sing their own praises, to engage in amorous contests or bitter rivalries, and to advertise the advantages of their specific natural attributes. For instance, arguing against adversaries that rashly consider mountains 'barren, rude, and voide of all delight' (7, 72), Malvern Hill – importing the principle of monarchical supremacy into the natural sphere – 'stoutlie [maintains] / Gainst Forrests, Valleys, Fields, Groves, Rivers, Pasture, Plaine, / And all their flatter kinds . . . / The Mountaine is the King: and he it is alone / Above the other soyles that Nature doth in-throne' (83–8).

This curious poetic fusion of history and geography begs to be linked to the tradition of geographical writing considered above. Alastair Fowler noted in 1984 that 'Drayton's poem has to be understood in relation to antiquarian topographical research: it was the age of Leland, Camden, Norden, Selden, and William Burton'.[14] Responding to this appeal some recent critics have explicitly reinserted the poem into the larger cultural context of which it was so obviously a part, the chorographical project of Elizabethan and Jacobean times. This connection was obvious to contemporaries: among the geographical guidebooks

John Taylor, the 'water poet', consulted on his extensive travels throughout England were the works of 'learned *Camden, Speed,* and *Hollinshead*' as well as '*Draytons* painfull *Polyolbion*'.[15] Helgerson has argued that *Poly-Olbion* continues and intensifies the anti-monarchical ideology of national chorography: with *Poly-Olbion* 'Drayton and his collaborators make known both their antipathy to Stuart absolutism and their allegiance to a rival source of authority'[16] – that is, to the land of the nation itself which Drayton's poem lends an eloquent voice, or rather, which emerges from the text as a deluge of different voices. McRae has usefully supplemented this analysis by pointing out that there is, nevertheless, a significant difference between *Poly-Olbion* and the English chorographical tradition. Calling the poem 'as much a troubled reaction as an enthusiastic contribution to the programme of Camden', McRae insists that Drayton's attention to national space 'consistently fails even to recognize the Jacobean gentry and their proprietary attachment to the land.'[17] Instead, Drayton idealizes rural life and foregrounds idyllic scenes of original splendour to etch a national mythology into the natural features of the land.

Exclusively animated by the spirits of the natural landscape and purged of all overt signs relating it to a hierarchical social order or a distinct pattern of settlement, the national space described here relies heavily on the triumph of the cartographic paradigm. Indeed, the text of *Poly-Olbion*, a poetic representation of landscape filtered through the visual powers of the invisible Muse, is unthinkable without the conceptual precedent of the map: the gaze that guides our own view as readers, that acts as an intermediary agent between land and text, is identical with the view of the modern cartographer. Early on in the poem, 'smooth-brow'd' (3, 122) Salisbury Plain defends its wide open space against 'barb'rous woods' (111) in whose 'darke and sleepie shades' hang 'mists and rotten fogs / . . . in the gloomie thicks, and make unstedfast bogs, / By dropping from the boughs, the o're-growen trees among, / With Caterpillars kells, and duskie cobwebs hong' (117–20). Dark and impenetrable woods block the unrestricted view a map provides but 'upon the goodlie *Plaines*' the light of the sun spreads into every corner of the terrain, illuminating 'this upper world' (128) with the glaring lucidity of its 'farre-shooting sight' (130).

Such passages amplify the distinction alluded to above in Norden's *Intended Guyde* and emphasize the indebtedness of Drayton's poetic construction of national space to the penetrating clarity of the surveyor's view. Indeed, the poem frequently draws on the technical vocabulary of the land surveyor to describe the activity of the peripatetic Muse: she 'measures out this Plaine; and then survayes those groves' (3, 348), she 'take[s] a perfect view / Of all the wandring Streames' (5, 90–1), she 'look[s] from aloft' to '[survey] coy *Severns* course' (7, 3–4), she 'survayeth' the land alternatively with an 'amorous' (8, 3) or an 'ambitious' (9, 2) eye, she even 'survay[s] every part' of the Welsh mountain *Dyffren Cluyd* 'from foote up to th[e] head', celebrating the mountain's 'full and youthfull breasts, which in their meadowy pride, / Are brancht with rivery veines, Meander-like that glide' (10, 92–4). The anthropomorphism of the latter passage, reappearing throughout the poem, seems to take literally the frequent claim in cartographic works that land – both naturally (as an island) and politically (as a kingdom) – appears to assume the form of a human body, thus requiring the cartographer to 'dissect' and lay open its geographical 'anatomy'.

Commenting on what he sees as a parallel in the internal composition of the *Britannia* and of *Poly-Olbion*, Helgerson has described Drayton's Muse as a poetic analogue and conceptual equivalent of the travelling chorographer moving through space: both the imaginative poetic progress through the land and Camden's physical perambulation of Britain represent individual and nation in a 'mutually enabling relation to one another'.[18] The similarity of their mediating function indeed suggests a structural identity of poetic Muse and antiquarian chorographer but their relation to the land rests neither on a shared intimacy of direct contact between the traveller and the space described, nor on the operational uncertainty of the voyage into unknown territory. Rather, their perception of national space is predicated on the absolutist claim of the map: like Camden and the county chorographers in his wake, Drayton's Muse views the land from the lofty position enabled by cartographic representation, she inspects national space not as the 'raw material' of landscape but as an image already filtered, already subjected to the scrutiny of the surveying eye. The Muse appears entirely as a cartographic fiction: the space she narrates will be fully exposed in the act of representation; the surface of the land, free of hermeneutic blanks, yields completely to her observant eye; and her travelling vocabulary owes its idiom of movement and action entirely to the finger moving across the map. It is important in this context that the Muse never enters the realm of the visible herself; like a 'point' in Euclid she remains elusive and insubstantial throughout the poem. Her 'travels' are an effect of the map, not of any direct spatial experience; her realm of movement is the liminal sphere created cartographically between *lived* and *conceived* spaces. In the same sense that the implied chorographer of the *Britannia*, equipped with preconceived principles of classification, resembles not so much a genuine traveller but a surveyor viewing landscape through a textual theodolite, the Muse appears as the prosopopoeia of the cartographic gaze, as the poetic abstraction of the visual synthesis offered by the geometric scale-map.

Criticism of *Poly-Olbion* has frequently concentrated on what appears to be Drayton's central objective, to offer up an image of Britain's geographical variety in the form of a unified and harmonious poetic whole.[19] If a further function of the Muse could then be described as the attempt to impose structural unity on what is experienced as a fragmented world, to rediscover an essential coherence in a geographical space marked by quarrelling rivers and warring mountains, this textual attempt at harmonization finds its most prominent complementary visual feature in the inclusion of 30 maps that preface the individual songs of the poem (Figure 8.1). These are curious maps, to say the least, but although many critics have noted their indebtedness to Saxton's atlas few have bothered to enquire about the nature of this derivative transfer.[20] In each case the terrain depicted corresponds

Figure 8.1   Michael Drayton: introductory map to song 15, containing the river marriage (1612).

roughly to the poetic description which follows, but the topographical information on offer is radically reduced: disregarding signs of human settlement such as towns or villages they focus almost exclusively on rivers, each graphically animated by its personal nymph, and the selection of toponyms spread over the cartographic surface follows no recognizable principle of order (other than the need to cross-reference the place-names mentioned in the poem). Compared to the maps of Saxton and Speed, the most obvious difference is their complete unwillingness to present landscape as a product of human civilization, as a cultural space shaped by the continuous presence of human society. These maps utterly ignore traces of the land's inhabitants; people are merely the accidental product of an environment that precedes all human interference with landscape.[21] As in the text of the poem itself, this discursive erasure does not stop short of the gentry. Ignoring the current order of property inscribed in the space of the nation, Drayton's England is exclusively inhabited by mythical creatures growing out of rivers or occupying the solitary hilltops of an ancient immutable landscape.

Lacking all further ornamental framework, these maps hover awkwardly between their referential attachment to a real landscape and the Edenic idealization of cornucopian abundance. What appears as the most striking aspect of the maps in *Poly-Olbion* is their mythic quality: the landscapes they chart seem utterly removed from the political territory registered in earlier cartographic work. One immediate indication of this discontinuity is the missing division into counties. Drayton's maps (though produced by the same engraver as Camden's[22]) circumscribe vaguely defined geographical regions, not administrative units. In semiotic terms, Drayton's maps consciously verge on the symbolic, for rather than suggesting a mimetic relationship between cartographic sign and physical landscape – as Speed's and Camden's 'iconic' maps attempt to do[23] – they refer us back to a fabulous mythology claiming to represent the national essence of England and Wales. In fact, the maps in *Poly-Olbion* are hardly recognizable *as* maps, let alone as maps of Britain: the featureless terrain resists even the most common cartographic impulse to focus on borders, and their geographic information hardly offers more than a bare 'summary' of the most prominent landscape elements, rivers and mountains. Their central function, I suggest, is to imagine the link with the subsequent poetic excursions across their surface as the implicit narrative of national purity which constructs the smoothness of the land's temporal transit from mythic prehistory (map) through the events leading up to the present (poem). The undisturbed constancy of this parallel move through space and time is further implied by the iconography of the maps' river network which features curving streams meandering wave-like across the terrain, like the pulsating veins of the national body, or the fertile roots implanted in the imperial garden.

But while the conscious symbolism of these maps may sustain Drayton's dream of an original historical purity of landscape recognizable in the very lie of the land, it fails to hide the tension implicit in *Poly-Olbion*'s multi-layered representational work, a tension that threatens to obliterate his project of national synthesis based on cartographic homogenization. If the maps are attempts to capture in mythological imagery an eternal and original truth about land, parading before the viewer a pictorial version of Britain's divinely ordained geography, they also admit their uselessness as a means of social and political orientation in the present. For the Britain of Drayton's 30 songs, as opposed to the Britain of his maps, is unable to continue the timeless mythology of its cartographic 'prefaces'. Instead, it engages in a narrative of constant historical change. At the same time as this symbolic cartography offers to depoliticize Britain – an attempt that is itself already reliant on a cartographic discourse firmly locked in a contest over the accurate representation of the social space of the nation – the text of the poem affirms the historical reality of continuous political factionalism. In reducing to a mere sporting contest the bitter rivalries between the constituent parts of its personified landscape, the nation *Poly-Olbion* imagines poetically may repeat the cartographic dream of an original spatial harmony, capable of transcending all social difference, but hardly serves as a model for a nation shaped by waves of invasions, or as the image of a society in transition.

The first 18 songs of *Poly-Olbion* are accompanied by a third discursive effort to make sense of Britain's topography, a historical commentary on Drayton's poem supplied by the foremost antiquary of the generation after Camden, John Selden. The relation of these notes to the poem reproduces the relationship between history and geography in the poem itself, where the temporal is equally treated as little more than a loose collection of footnotes to the spatial. Their purpose was to supply additional historical information and to draw the poet back to earth from occasional flights of fancy: 'What the Verse oft, with allusion, as supposing a full knowing Reader, lets slip; or in winding steps of Personating Fictions (as some times) so infolds, that suddaine conceipt cannot abstract a Forme of the clothed Truth, I have, as I might, *Illustrated*.'[24] This overt separation between *story* and *history* on the pages of *Poly-Olbion*, bearing in mind that the categorical distinction between truth and fiction was still a fragile construction when the poem was being written,[25] serves to further disrupt any overall sense of spatial coherence, adding to the conflict between the visual and the verbal, between the poetic and cartographic trajectories of the book. For in its triadic structure of map, poem and history, *Poly-Olbion* seems less to unite disparate areas of knowledge in a comprehensive account than to underline their dissimilar nature, presenting to contemporary readers an image not of eternal union but of internal fragmentation, affecting nation and knowledge alike. What the introductory maps attempt to cover up, the existence of discord and dissent, is re-imported into the poem through the explicit segregation, and mutual incompatibility, of the discursive communities assembled on its pages.

Drayton's topographical descriptions, specifically in the latter songs of *Poly-Olbion*, are frequently interlaced with extensive historical catalogues listing England's rulers (17), its military heroes (18), its famous sea captains (19) and saints (24), along with lengthy inventories of natural attributes cataloguing England's birds (13), its fish and fowl (25–6), and so on. Such exhaustive lists affirm the proximity of *Poly-Olbion* to the textual conventions of the chorographical tradition, but also serve to indicate a vital functional difference between the respective spatial imagination of Drayton and Spenser, a difference best formulated, I suggest, as the opposition between static and dynamic visions, between still life and performance. Drayton's evocation of England's musical tradition, the opening gambit of England's bid to claim the island of Lundy from the Welsh in song 4, allows an initial comparison:

> The trembling Lute some touch, some strain the Violl best
> In sets which there were seene, the musick wondrous choice:
> Some likewise there affect the Gamba with the voice,
> To shew that *England* could varietie afford.
> Some that delight to touch the sterner wyerie Chord,
> The Cythron, the Pandore, and the Theorbo strike:
> The Gittern and the Kit the wandring Fidlers like.
> So were there some againe, in this their learned strife
> Loud instruments that lov'd; the Cornet and the Phife,
> The Hoboy, Sagbut deepe, Recorder, and the Flute:
> Even from the shrillest Shawme unto the Cornamute.
> Some blow the Bagpipe up, that plaies the Country-round:
> The Taber and the Pipe, some take delight to sound.

(4, 351–68)

Engaged in a contest over the cultural ownership of space, musical instruments line up almost in explicit analogy to a battle formation, ready to triumph over an opposing orchestral side. In a rising crescendo, deliberate emphasis is placed on their martial qualities: 'loud', 'sterner', 'shrillest'. Their function is to translate an essential Englishness into the vibrancy of a musical pageant, '[t]o shew that *England* could varietie afford', but the point I wish to stress about this passage is that Drayton's encyclopaedic style works in fact to disguise the performative aspect of the activities described in the image; the impression is not of musicians in action but of an exhaustive list specifying all manner of musical instruments available to contemporaries. Consider the contrast to this passage from *The Faerie Queene* which describes the musical sensation experienced by Arthur and Guyon, the knight of temperance, on their approach to the Bower of Bliss in the final canto of Book 2:

> Eftsoones they heard a most melodious sound,
>   Of all that mote delight a daintie eare,
>   Such as attonce might not on liuing ground,
>   Saue in this Paradise, be heard elswhere:
>   Right hard it was, for wight, which did it heare,
>   To read, what manner musicke that mote bee:
>   For all that pleasing is to liuing eare,
>   Was there consorted in one harmonee,
> Birdes, voyces, instruments, windes, waters, all agree.

> The ioyous birdes shrouded in chearefull shade,
>   Their notes vnto the voyce attempred sweet;
>   Th'Angelicall soft trembling voyces made
>   To th'instruments diuine respondence meet:
>   The siluer sounding instruments did meet
>   With the base murmure of the waters fall:
>   The waters fall with difference discreet,
>   Now soft, now loud, vnto the wind did call:
> The gentle warbling wind low answered to all.

> > (II.xii.70–1)

The striking self-reflexivity of this passage is immediately apparent. Spenser's text literally enacts the theme of musical polyphony: the asyndetic stringing together of 'Birdes, voyces, instruments, windes, waters' in the final line of stanza 70, echoed by an explicitly musical arrangement of repetition and variation in the ensuing stanza, presents natural and human sound effects 'consorted in one harmonee'; words like 'trembling', 'murmure', 'warbling' revel in an unashamed show of melodious onomatopoeia. If Spenser's dynamic polyphony effortlessly manages to imitate music, Drayton's static catalogue labours to enumerate the individual components of a semantic field.

Of course, these respective treatments of the theme of music may be too divergent in their poetic intent to allow a meaningful comparison: Spenser concentrates not on a parade of instruments but on their musical effect, he foregrounds the experience of exposure to sound, not its technical source. But this is precisely the point. Drayton's poetry persistently aims to offer a full and final description of its chosen theme, it works meticulously (like a map) through a complete and immobile pictorial enclosure. Spenser hardly ever displays a similar degree of descriptive fixity. If *The Faerie Queene* never fails to require its readers to enact their own meaning, *Poly-Olbion* nowhere allows the dangerously undefined space of interpretative activity.

My comparison between these musical passages is not intended to deride *Poly-Olbion* for what might well seem like the poverty of its aesthetic imagination but to clarify an internal contrast between Drayton's and Spenser's poetic techniques; a contrast I find reproduced in both epics' conception of geographical space. Perhaps its most overt manifestation is provided by the alternative poetic treatments of a prominent spatial theme, the river marriages recounted with slight alterations in Book 4 of *The Faerie Queene* and in the 15th song of *Poly-Olbion*. For both poets the fluvial image of conjugal union figures as a model of harmony created out of difference, yet each chooses a different conceptual form for its textual realization. Drayton imagines the marriage between male Tame and female Isis as the union of two rivers 'both so lovely . . . , that knowledge scarce can tell, / For feature whether hee, or beautie shee excell' (99–100). From the forging together of their individual qualities, new life will spring: 'Betwixt your beautious selves', the muses announce to Isis, 'you shall beget a Sonne, / That when your lives shall end, in him shall be begunne' (15, 103–4). A giant wedding feast accompanies the birth of the river 'Tamesis' (16, 2), or simply Thames, in which the surrounding landscape happily participates. The vale of Alsbury spreads flowers along the path, the old hill of Chiltern offers paternal advice to his son Tame, and the local rivers come rushing on to contribute their share to the successful ceremony: the Windrush is busy preparing the feast; the Cherwell delivers a glowing speech in praise of rivers; 'the *Oxonian Ouze*', in charge of public relations, is sent away '[t]o publish that great day in mightie *Neptunes* Hall, / That all the Sea-gods there might keep it festivall' (15, 63–4).

Rivers are part of an endless regenerative cycle; their flow of water is continuous. The confluence of two streams, an occurrence of unceasing regularity, can be meaningfully described in terms of its spatial location but hardly as an event in time. It may thus seem odd that Drayton chooses expressly to temporalize his vision: both rivers meet on a 'day of Mariage' (15, 2); on her first anxious approach Isis '[c]omes tripping with delight' (15, 67), then 'hasts more speedily' (15, 119) past Oxford to avoid delay; after Cherwell's speech 'the wedding ends, and brake up all the Showe' (15, 281). Why this need to build up the explicitly spatial event of a river marriage to a temporal climax? The reason, I would like to suggest, is the unavailability of space as a flexible category of imaginative description: Drayton's English geography is fixed and immobile, a complete act of faith in the topographical map. His landscape may be animated by babbling brooks and mumbling mountains but never breaks free of its cartographic confinement. Indeed, throughout *Poly-Olbion* Drayton's prosopopoeia always stops half way. His rivers are never fully anthropomorphized; they may have an eloquent, even critical, voice but never once leave their allocated beds. Like the Humber, the Trent or the Severn, the Thames remains forever a solitary river and will only meet its continental counterparts, '[t]he *Skeld*, the goodly *Mose*, the rich and Viny *Rheine*', once it loses itself completely in '*Neptunes* watry Plaine' (15, 109–10). Spenser, in a canto from *The Faerie Queene* that may well have served Drayton as a model, does not display the same geographic fidelity. His river marriage (taking place further down the Thames estuary, at the point where it receives the Medway) is imagined as a vast triumphal procession, a fully worked through prosopopoeic spectacle. The highly formalized wedding ceremony is attended by rivers from around the globe: the Nile, the Rhône, the Danube, Ganges, Euphrates and Tigris, the Rhine and the Tiber, even '[r]ich Oranochy, though but knowen late' (VI.xi.21), all make their appearance at the solemn feast 'in honour of the spousalls, which then were / Betwixt the *Medway* and the *Thames* agreed' (8).

Spenser's procession is doubly sequential in nature. Emanating from, and returning to, a sovereign centre, it also follows a determined order of historical time:

Led by the elemental gods Neptune and Amphitrite and their brood, and followed first by the aquatic founders of nations – Albion and Inachus among them – and by Ocean and Tethys who 'both sea and land possest' and then by the famous rivers associated with successive periods of human history, the procession brings the flow of

history forward in one continuous, unbroken stream – one water
with a common source and outlet.[26]

Beginning with Neptune and finding its climax in the union of Thames
and Medway, the logic of the cycle then reverses this sequence to end
where it started, in the timeless oceanic source of all human life on
earth. In its internal structure, the procession enacts a strictly hierarchical order: smaller rivers are attendant upon larger streams, Ireland's
waterways follow England's in 'duefull seruice' (44). Punning on the
semantic overlap of tribute and tributary – feudal gesture and hydro-
graphic label – Spenser charges the entire procession with high political energy: 'the river marriage presents a fantasy of power couched
largely in feudal, chivalric terms, the language of tribute.'[27] Enabling
this powerful founding myth is Spenser's instrumentalization of an
ancient political idea which conceives of the land as coextensive with
the body of the ruler: the event of the fluvial reunion situates the realm
both in historical time and in the social order of the present, imagining a triumphal procession where 'mightie *Albion*, father of the bold /
And warlike people, which the *Britaine* Islands hold' (15) mixes with
'[a]ll little Riuers . . . [who] owe vassallage' (29) to their sovereign lord,
the mighty *Thamis*.

Based on the examples of Leland and Camden,[28] Spenser's use of the
river marriage is as obviously self-reflexive as it is painstakingly topical;
his nuptial river procession is 'the place where world and word meet.'[29]
'[S]torming Humber' (30), 'stately Seuerne' (30), 'Stoure with terrible
aspect' (32), all openly parade speculative etymologies, while a
distraught poet – in a phrasing reminiscent of 1 Cor. 13 – laments
that 'not if an hundred tongues to tell, / And hundred mouthes, and
voice of brasse I had, / And endlesse memorie, that mote excell, / In
order as they came, could I recount them well' (9). A few stanzas
further on, the centrality claimed by the issue of representation is
again affirmed in the form of a reflection on the poet's difficulties to
accommodate all participating rivers into the limited textual space
available for description:

> But what doe I their names seeke to reherse,
>   Which all the world have with their issue fild?
>   How can they all in this so narrow verse
>   Contayned be, and in small compasse hild?
>   Let them record them that are better skild,
>   And know the moniments of passed times:

> Onely what needeth, shall be here fulfild,
> T'expresse some part of that great equipage,
> Which from great *Neptune* do deriue their parentage.
>
> (IV.xi.17)

The easy answer to the question posed in lines 3–4 might of course be to recommend the representational short cut of the topographical map; and in the central section of the pageant, where the English and Irish rivers join the proceedings, the whole canto indeed momentarily appears like 'a meditation on a map of the British Isles'.[30] But, as Berger further notes, the narrative presentation of local rivers as part of a cosmological (not chorographical) procession actually 'leaves maps far behind': by forcing rivers into the strict hierarchical order of the rigid cultural form offered by a triumphal ceremony, Spenser's 'principle of meaning and visualization is rhetorical, not cartographic'.[31]

Or, one might argue, the map was quite simply not available conceptually to structure this imaginative river pageant. Biographical evidence could be enlisted to support this point. Commenting, in an early letter to Gabriel Harvey, on the idea for a poetic work that may well have been the origin of the river marriage of Book 4, he cites as his sources not the maps of Christopher Saxton (which were already available in print) but, significantly in my view, Harrison's *Description*:

> I minde shortely at conuenient leysure, to sette forth a Booke . . . whyche I entitle, *Epithalamion Thamesis*, whyche Booke I dare vndertake wil be very profitable for the knowledge, and rare for the Inuention, and manner of handling. For in setting forth the marriage of the Thames: I shew his first beginning, and offspring, and all the Countrey, that he passeth thorough, and also describe all the Riuers throughout Englande, whyche came to this Wedding, and their right names, and right passage, &c. A worke beleeue me, of much labour, wherin notwithstanding Master *Holinshed* hath much furthered and aduantaged me, who therein hath bestowed singular paines, in searching oute their first heades, and sourses: and also in tracing, and dogging out all their Course, til they fall into the Sea.[32]

When Spenser wrote this (before 1580), Camden's *Britannia* had of course not yet been published but the explicit link with Harrison's *Description* (mistakenly attributed to the overall editor of the chronicles, Raphael Holinshed), rather than with Saxton's maps, suggests a stronger influence than the mere accident of textual availability. The form of the itinerary characteristic of Harrison's river journey across the nation shares a performative quality with Spenser's spatial imagination that is all but submerged by the static descriptive patterns of the *Britannia* and its poetic successor, *Poly-Olbion* (Spenser's later praise for Camden in *The Ruines of Time* notwithstanding[33]). Thus, in *The Faerie Queene*, England's rivers only rise to their full meaning as participants in the active performance of a ritual marriage, not as lines on a map, and as the chorus of Spenser's 1596 marriage poem *Prothalamion* reminds us, the congruence of art and geography is the enabling condition of an idealized, free-flowing river landscape where aesthetic surplus is never merely a cartographic imitation. Rather, poetic and fluvial progression are imagined as merging in the steady parallel flow of water and verse: 'Sweet *Themmes* runne softly, till I end my Song.'[34]

These different river mythologies point to a conceptual design shaping the spatial imagination of *The Faerie Queene* which consistently undermines the impulse to pull out a map and ground the action of Spenser's romance plot in geographical space. In a sense, the poem quite deliberately charts, in the novelist Michael Ondaatje's phrase, an 'earth that had no maps'.[35] An obvious instance of this anti-cartographic trace is the voyage of Arthur and Guyon to Acrasia's lustful bower in the closing canto of Book 2 (a journey that leads up to their musical reception quoted above). The complexities of their approach to the Bower of Bliss has troubled criticism. In an incautious moment, one commentator suggests in an introductory textbook on *The Faerie Queene* that students of the poem should draw a map of the eventful journey to facilitate understanding of Spenser's allegorical scheme.[36] This mapping impulse may initially seem a plausible pedagogical device in teaching the section of *The Faerie Queene* which has long been recognized as 'the most landscaped part of the poem'.[37] It invites, however, a doubly revealing mistake. It is hard to imagine practically how a map could be drawn of a perpetually changing landscape that includes such topographically unstable elements as wandering islands which 'seeming now and than, / Are not firme lande, nor any certein wonne, / But straggling plots, which to and fro do ronne, / In the wide waters' (II.xii.11). And it is equally difficult to see how a map could be of aid conceptually to a Spenserian knight (and thus, by hermeneutic implication, to Spenser's readers) who need to move through an incessantly moralized landscape where the right path of virtue will only be discovered after the endurance of extensive moral trials.

Spenser's allegorical landscapes are realms of constant illusion and deceit, posing challenges which the characters first need to successfully confront before they can safely move on. The degree to which Fairyland consistently evades the cartographic paradigm is perhaps best illustrated by Britomart's predicament in the book of chastity, when she embarks on her perilous quest '[w]ithouten compasse, or withouten card' (III.ii.7). Rather than offering the full transparency and safe orientation of a map, the geography of Fairyland is composed of standardized elements which need to be read accurately both by passing knights and learning readers: castles are divided into houses of virtue and anti-virtue; forests are 'always a place of mystery, full of unseen dangers and occasional pleasant surprises';[38] caves may be dens of wild beasts or places of temptation; water – fountains, rivers, lakes – tends to be associated with the spiritual force of baptism but may harbour either chaste or unchaste nymphs. Plains alone appear to be relatively free of hidden moral perils, acting most frequently as the stage for chivalric combat or chance encounters between travelling knights. But even in unrestrained, open space, identities are more often misread than not. Mastering the complex road system of *The Faerie Queene* is an equally double-edged task. Ten cantos into the legend of holiness, the Redcrosse knight is finally told what he painfully found out for himself, that he should shun the 'broad high way' in favour of 'the narrow path' (I.x.10); but this he cannot, indeed *must* not know prior to his first major adventure – if the poem is to fulfil its didactic purpose.

Thus, his initial challenge in the first canto of Book I is duly presented as a lesson in perception.[39] Forced to seek shelter from 'an hideous storme of raine', Redcrosse and his companion Una enter into 'a shadie groue' (I.i.8–9). What appears at first sight like a secure network of 'pathes and alleies wide' (9) quickly metamorphoses into a confusing maze: 'wander[ing] too and fro in wayes vnknowne, / Furthest from end then, when they neerest weene, / . . . So many pathes, so many turnings seene, / That which of them to take, in diuerse doubt they been' (10). Their aimless wandering only announces, by way of etymology, a mortal danger hiding in the centre of the forest, the dragon 'Errour'.[40] Since the beast is as much a physical threat as a mental condition, moral and geographical orientation will only set in after the dragon has been slain by the knight, and Redcrosse's initial choice of the path 'that beaten *seemd* most bare' (11, my italics) as a potential exit proves, almost by necessity, *erroneous*. Bragging to his anxious companion Una, in true male fashion, that '[v]ertue giues her selfe light, through darknesse for to wade' (12) he ploughs on, only to stumble straight into Errour's den. The promise of full illumination is ironically reduced to a 'litle glooming light, much like a shade' (14) – an accidental reflection from his armour which keeps himself in the dark but wakes the dragon – yet with the support of Una's un*erring* faith, he successfully manages to defeat Errour (and thus to correct his own near fatal error). It is only after this heroic deed that the landscape opens up to his gaze; and the narrow path that before '*seemd* most bare' now turns into the path 'that beaten *was* most plaine' (28, my italics), to lead them safely out of danger – until they encounter the next spatial challenge.

The example shows that in *The Faerie Queene* physical space has to be mastered first before it can become fully transparent to any knight passing through it; Fairyland is above all a moral testing ground which is not accessible from the privileged position afforded by the topographical map. The attempt to situate the action of Spenser's poem in historical and geographical space fails to acknowledge this experimental status of *The Faerie Queene*'s setting.[41] The crucial point is not how much 'real' geography finds its way into the poem (there is a good deal) or according to what schemes the fictional landscape is absorbed into a higher cosmological order; rather, the poem undertakes to question the epistemological significance of space as such, and to describe its power to define the existential state of the fictional characters moving through it. Spenser's landscapes are never closed, purged of dangers, or shut off from the intervention of threatening 'others':[42] Fairyland is as much a romantic setting as it is one endless physical and moral struggle. In this aesthetic universe, where place is always also a moral and political condition, cartography borders on a conceptual impossibility: if knowledge is indeed the ideal end result of a lengthy process of moral education (as suggested in Spenser's letter to Ralegh[43]), not the point of departure, then maps leap straight into an intellectual condition which the troubled knights of *The Faerie Queene* can only ever hope to achieve at the very end of their own inner journey.[44]

If a map is thus an inconceivable short cut for a Spenserian knight, which may facilitate spatial displacement but only at the cost of dangerously undermining the overall disciplinary project of the poem, this anti-mapping attitude, I suggest, is not just a quaint poetic device but is linked to a contemporary discussion over the relative merits of modern cartography. To illustrate this debate I need to briefly return to the surveying manuals examined in Chapter 2. One of the many uncertainties surrounding new methods of measuring land was the realization that maps were not entirely faithful to the specific natural conditions of the land they depicted. Cartography was always partial and provisional, and it deliberately needed to 'arrange' the ground for the purposes of visual representation; maps forced the irregularities of landscape both onto a flat piece of paper and into a rigid geometric grid. General acceptance of this labour of cartographic abstraction was less widespread than might be assumed. According to Radolph Agas, a practising surveyor, the map's effacement of the ground's true surface 'hath perswaded many wise and excellent persons, to doubt whether there be perfection in mapping of landes and tenements for surueigh', for in light of 'the vneuenes of the groundes, . . . their great difference, in hill and dale, from a leuell superficies', surveyors 'are necessarily compelled to put downe [their] practise vpon bookes that are leuel and smooth'.[45] Aaron Rathborne even spotted considerable ignorance concerning the complexities of cartographic projection among his own colleagues. Some 'plaine plaine [*sic*] Tablemen', he fumed, would 'at an instant . . . conuert the highest mountains to plaine and leuel grounds, pressing them downe, and inforcing them on a Plaine sheete of paper to lye leuell with the rest'.[46]

I find a striking echo of these anxieties about the (im)possibility of cartographic 'accuracy' in an episode from *The Faerie Queene* which is traditionally read as an allegorical warning about the dangers of popular democracy. In Book 5, Canto 2, the stern justice of the titular hero Artegall is visited upon a plebeian giant who takes up his position at the liminal point of the sea coast and proposes to a cheering populace that, since the world has run so much out of its original proportion, he would now weigh it anew with his huge pair of scales – seemingly an attribute of justice – and thus restore an original state of perfect equality.[47] After a lengthy discussion, which duly ends with the giant being thrown over a cliff by Talus, Artegall exposes this version of justice as dangerous egalitarianism which threatens to undermine the divinely sanctioned hierarchy of earthly affairs. The relevance of this scene for my purposes is the nature of the giant's reformative vision which focuses not on material goods but on the shape of the land. In order to recover the world's original justice the giant proposes to 'throw down these mountains hie / And make them levell with the lowly plaine: / These towring rocks, which reach vnto the skie, / I will thrust downe into the deepest maine, / And as they were, them equalize againe' (V.ii.38). Though hardly intended as a conscious allegory for the act of mapping, the levelling giant here suggestively assumes the guise of a transgressive cartographer. In forcing mountains to lie 'levell with the lowly plaine', he verbally closes ranks with Rathborne's dilettantish surveyors who equally flatten out landscape by 'pressing down' the highest mountains 'to lye leuell with the rest.' In Book 5 the natural irregularities of the ground's surface, which the egalitarian giant wrongfully attempts to correct, are conflated with a divinely authorized political landscape reflecting the principle of social order and the distribution of power:

> The hils doe not the lowly dales disdaine;
> The dales doe not the lofty hils enuy.
> He maketh Kings to sit in souerainty;
> He maketh subiects to their powre obay;
> He pulleth downe, he setteth vp on hy;
> He giues to this, from that he takes away.
>
> (41)

Ever since their rebellion against Jove, related by Ovid and others, giants are a traditional motif of transgression. In Book V, the thrust of their activities is significantly reversed: when the giants were first reported to 'caelestial Thrones affect [and] to the skies congested Mountaynes reare',[48] their upward movement sought to usurp an elevated hierarchical centre. Spenser's mock Atlas figure translates this vertical challenge to authoritarian monarchical claims into a horizontal redistribution of landscape elements, an activity that recalls the suggestion of political equality fostered by the cartographic grid image of the national community.[49] Physical landscape, still invested (by Spenser) with the semiotic potential to stand for the body of the monarch, is a direct mirror of political power; unlawful interference with the spatial configuration of the nation is registered as a rebellious act. Smoothing out an irregular surface, if symbolically in conscious literary allegory or materially on a plane piece of paper, contains the suggestion of similar levelling actions in the political sphere. Since the space of the nation is a fragile figure, produced by a process of narration that consistently includes and acknowledges, albeit with destructive intent, the existence of disruptive others – like giants, dragons and savages – it demands its constant reiteration through active performance, rather than its mechanical pictorial reproduction through maps. Unlike *Poly-Olbion*, which offers cartography as a fully sufficient generative matrix for the idea of the nation, the space of *The Faerie Queene* is dependent on its prior articulation through the combined collective effort of the national community constitutive of all its individual members.

This distinction, I would like finally to suggest, recalls terms used by Homi Bhabha in his influential analysis of modern nationalist discourse. According to Bhabha, 'the contested conceptual territory' of the national idea is evidence of two opposing but mutually enabling strategies, first, a 'nationalist pedagogy, giving the discourse an authority that is based on the pre-given or constituted historical origin *in the past*', and second, 'a process of signification that must erase any prior or originary presence of the nation-people to demonstrate the prodigious, living principles of the people as a contemporaneity: as the sign of the *present* through which national life is redeemed and iterated as a reproductive process.'[50] Within the parameters of the former strategy, people constitute the 'objects' of the national discourse, within those of the latter, its 'subjects'. If the terms of Bhabha's critical project – developed in response to the modern postcolonial experience of exile and geographic dispersal – may be accorded relevance for the early modern period, I suggest that Drayton's narration of national space embraces predominantly 'the continuist, accumulative temporality of the pedagogical' while Spenser's vision of the nation as Fairyland is principally owed to 'the repetitious, recursive strategy of the performative'.[51] Between the didactic project of teaching the nation as a preconceived image of structural coherence stretching back through linear time, and its constantly reiterated production in the present by a collective social and moral effort, lies the contested terrain of an early modern poetics of national space. If the pedagogical short cut of the map no longer caused the same anxieties when Drayton announced that his cartographically inclined 'industrious Muse great *Britaine* forth shall bring' (1, 65), his poetic map of the nation signals two decisive political changes, first, the decline of an ancient constitutional idea that saw land and ruler, space and power, yoked together in a seamless semantic continuum,[52] and second, the replacement of a concept of national integrity based (albeit hesitantly) on inclusion, by an idea of collective identity reliant on strategies of exclusion. For both developments the success of the cartographic paradigm proved crucial, and it is the latter change to which the narrative construction of Irish space most directly bears witness.

9
Groundless Fictions: Writing Irish Space

Descriptive accounts of Ireland written parallel to the English choro-graphical project are rarely seen as being inspired by the same kind of antiquarian and patriotic impulse that sparked off the projects of Camden and his successors. Of the numerous Irish tracts written in late Tudor and early Stuart times, few if any conceive of Ireland as a national or regional entity with the same claim to spatial and political integrity as Britain or any of its individual counties. Generally, a contemporary 'description', 'view' or 'survey' of Ireland tends to be a pragmatic reform tract, containing a range of practical suggestions on how to pacify what was considered a rebellious and barbaric island.[1] The single exception to this rule is perhaps a rarely noted manuscript by John Norden, entit-led simply *A Description of Ireland*, which Norden prepared for the earl of Salisbury, then Lord Treasurer, in 1610:[2] as if intended as a late addi-tion to his *Speculum*, a map of Ireland (based on Boazio) is accompanied by statistical material on Irish towns (including grid references), on garrisons, castles and forts, on earldoms and seignories, provinces, coun-ties, rivers and loughs, on distances to Spanish, French, English, Scot-tish towns and ports, and so on. Of all extant contemporary Irish tracts (of which I am aware) this interesting document – clearly a presenta-tion copy – comes closest in layout and intent to the conventions of an English county chorography.

That Norden's manuscript should remain the only excursion into the descriptive terrain so thoroughly explored in Britain testifies to the Janus-faced nature of contemporary Ireland. On the one hand evidence of the impulse to incorporate Ireland within a national framework, Norden's Irish chorography is also a clear indication of the undefined and confusing space it represented to English observers, a crux to which Norden's preface bears witness: worried about factual inconsistencies marring his work he asks Salisbury 'to have patience with the defectes, in regarde of the tedious conferring of so manie disagreyng plotts together, which can not be trulie reconciled by greatest care'. Not working from fresh surveys but relying on the work of others (it is unlikely that Norden ever set foot on Irish soil), his map seemed a wild distortion, a treacherous and unreliable patchwork:

> [B]y conference of seuerall plotte [the *Description*] would grow more and more confused. For that men differ muche among them selues in conceyuing thinges they seeme equallie to knowe, breeding rather new ambiguities then reforming the olde ... [N]ether my weake skills, nor greateste industrie ... can bringe this confused Plott to that righte shape and true proportion.[3]

Norden's manuscript thus repeats the circular pattern that emerged as a defining characteristic of contemporary Ireland's visual and textual configuration in the work of English topographers: just as its constitutional status oscillated between the political categories of kingdom and colony,[4] the assumption of Irish sameness could always produce an image of difference (and vice versa). The desire to apply the anatomical tools of chorographic dissection to the Irish territory, and to absorb it into the larger political framework of British nationhood, only affirms what motivated the project in the first place, the existence of an intractable otherness just across the Irish Sea.

In Chapters 3 and 6, I have traced similar patterns both in the debate over Ireland's political reform, where geographically inspired schemes to 'cleanse' Irish soil of its inherent barbarism instrumentalized the homogenizing logic of the surveying paradigm developed on the English estate, and in the changing shape and texture of Ireland's topographical image on the contemporary map, which visually negotiated its ambivalent relationship to the cartographic construction of a still imprecisely defined 'nation'. In this final chapter, I offer to explore the equally circular function of narrative in the cultural representation of Irish space.

To insist on a conceptual model that allows the appreciation of structural similarities while paying attention to the textual record of difference is not to disregard completely a dominant critical stance which insists on a deeply ingrained antagonism between each cultural realm. To be sure, the most powerful cultural narratives generated by the English desire to bring the entire island under political and military control, and thus to complete a conquest begun centuries before, emerge from two discursive fields which hardly justify a conceptual approach insisting on the notion of sameness: nomadism and savagery. In Paul Brown's definition, nomadism – or, in his term, 'masterlessness' – 'analyses wandering or unfixed and unsupervised elements located in the internal margins of civil society . . . [while] [s]avagism probes and categorises alien cultures on the external margins of civil power'.[5] Constituting categories of inversion and otherness, rather than terms of comparison, neither the construction of the Irish as aimless wanderers nor as uncivilized wild men appears to significantly challenge the oppressive and essentially one-directional nature of what has been identified as a self-authenticating structure: labelling as barbarian what is perceived as culturally different defines only the presumed civility of the speaker.[6] But the parallel construction of the Irish as internal rebels and external savages already implies some of the complexities structuring English perceptions of Ireland. As Andrew Hadfield has pointed out, where colonial and national discourses operate simultaneously, no simple separation of the domestic and the exotic suggests itself.[7] It is thus important not to evade the implications of the multiple cultural, institutional and geographical proximities between Britain and Ireland – the 'long history of contact'[8] between both islands – in favour of a monolithic conception of colonial power. When contemporary texts stress an 'essential' difference between Ireland and Britain, they often work only to conceal the inseparable links and points of contact generated by a common history.

Thus, if nomadism and savagery are accepted as the principal discursive parameters organizing the English perception of Ireland, I find the spatial concerns at the core of both discourses reminiscent of the semantic conflict over plan and itinerary which characterized the social space encompassed by the English chorographical project. The Irish nomad – imagined by Tudor historians as a descendant of Herodotus' Scythians – is fundamentally a spatial fiction: wandering with his cattle over the meadows and pastures of a potentially fruitful but sadly ill-treated land,[9] his unstructured itinerary fails to define space in terms identifiable by the classificatory systems of English antiquarianism. The space produced here is without purpose or structure, uncultivated and raw. Classical precedent helped shape this view. In Herodotus' *Histories* the Scythians occupy the geographical margins of a symmetrically organized *oikumene*, their world is available to representation only in terms of difference and deviation from the Greek norm.[10] Similarly, in the English discourse on Ireland, non-sedentarism is not considered as an accidental side-effect of the Irish agricultural practice of transhumance (the seasonal moving of livestock between different pastures) but constructed as the very opposite of English settlement patterns. Literary models of ancient geography are thus instrumental in turning the native Irish indiscriminately into aimlessly wandering vagrants or masterless men.

The proto-ethnographic concept of savagery implies no less a spatially structured discourse. By way of analogy with the agricultural sphere, Sir John Davies explained that for a successful political reform in Ireland

> the Husbandman must first breake the Land, before it bee made cape-able of good seede: and when it is thoroughly broken and manured, if he do not forthwith cast good seed into it, it will grow wilde againe, and beare nothing but Weeds. So a barbarous Country must be first broken by a warre, before it will be capeable of good Gouernment; and when it is fully subdued and conquered, if it bee not well planted and gouerned after the Conquest, it wil eft-soones return to the former Barbarisme.[11]

The military strategy which rationalized a ruthless war of attrition as a necessary precondition of political pacification is common to many Irish reform tracts, as is the popularity of the state-as-garden metaphor serving as a blueprint for political action.[12] What strikes me as most relevant in this passage is the nature of the link Davies establishes between people and land: in wild and archaic places like Ireland, no essential difference exists between uncultivated soil and its barbarous inhabitants. This metonymic identification grounds historical and cultural arguments about Irish savagery firmly in geographical space, and processes the relevant ethnographic information on Ireland's population through explicitly spatial paradigms.

This is perhaps nowhere more obvious than in the 1620 *Discourse of Ireland* written by Luke Gernon, a provincial English judge then resident in Limerick. This manuscript epistle addressed to a friend in England has become notorious for the disarming transparency of its oppressive male fantasies of rape: 'This Nymph of Ireland, is at all poynts like a yong wenche that hath the greene sicknes for want of occupying.'[13] The analogy between Ireland and a female body – giving a further semantic twist to Davies' reference to the violent 'Husband-man' necessary for Ireland's reform – is enabled by an explicitly cartographic conceit. 'Ireland shall be my theame', Gernon begins, but then refuses to follow a textbook geography which would require him to 'describe the clymat, the degrees, the scituation, the longitude, the latitude, the temperature, &c.' Such pragmatic information, readers are advised, may be found elsewhere: 'Go look in yor mapps'. Instead, Gernon opts for what he considers 'a more quaynt and genuine devise'. Vaguely recalling an image that may have been a copy of the virgo map of Europe in post-1588 editions of Münster's *Cosmography*,[14] he explains that

> [i]t was my chance once in a place, but I know not where, to see a map of Europe, and it was described in the lineam^{ts} of a naked woman, and upon the surface was a mapp of the countreyes. . . . In such a forme will I represent our Ireland, and yett, if my cunning fail me not, I will depaynt her more lively and more sensible to yor intelligence then if you had her in a table.[15]

This programmatic statement of intent is duly put into practice, producing such predictably misogynist analogies as the observation that feminized Ireland 'is very fayre of visage, and hath a smooth skinn of tender grasse', or that '[h]er breasts are round hillockes of milk-yeelding grasse, and that so fertile, that they contend wth the vallyes. And betwixt her leggs (for Ireland is full of havens), she hath an open harbor, but not much frequented.'[16]

In quoting these passages, my intention is not to revel in the pornographic voyeurism of Gernon's language but to draw attention to the overt instrumentalization of the cartographic paradigm as a means of political subjection. The difference between a mere 'table' and the feminized Ireland superimposed on it, as well as the explicit priority attached to the latter image regarding its expressive power, is a variation of the distinction I noted in Chapter 1 with regard to the two maps of Europe in Münster's cosmography – one 'accurate', one symbolic: the pragmatic 'table' – that is, the geometric scale-map – enables the rational analysis of Ireland's topographical and economic data but the country's true nature, its very essence, only becomes available in the 'quaynt and genuine devise' of the feminized cartographic allegory, being at once 'more lively and more sensible to yor intelligence'. While both maps offer substantially different information they are locked together in the mutually enabling relationship constitutive of colonial power: the first answers to the practicalities of conquest by supplying a detailed description of Ireland's physical geography, the object of military surveillance, the latter to the needs of ideological legitimation by organizing topographical data in explicit acknowledgment of a hierarchical system of gender relations that serves to naturalize a relationship of political domination.

Gernon's description clearly privileges the discursive image over the topographical information, and in light of the subsequent textual references to 'desolate' and 'destroyed' cities, to 'noysome, & stincking houses', or to 'the desolation wch cyvill rebellion hath procured',[17] it is possible to read his Irish virgo map as an instance of the powerful trope, ubiquitous in colonial discourse, which Louis Montrose has analysed with reference to the disturbing background cannibalism on Jan van der Straet's drawing *America*. Through the call for masculine control implicit in the formal relationship between the dominant image of the feminized land and the supporting rhetoric of destruction and waste, which mirrors the conflation of barbarism and femininity on *America*, the *Discourse of Ireland* employs in its textual structure 'a figure for the dynamic of gender and power in which the collective imagination of early modern Europe articulates its confrontation with alien cultures.'[18] Recent critical work has insisted on the frequency of such gendered tropes in images and texts emerging from the colonial encounter. The image on which Montrose bases his analysis, van der Straet's *America* (1575), which circulated widely across Europe in the form of a late sixteenth-century engraving by Theodor Galle, is undoubtedly among the best known visual instances of this trope. Staging the confrontation between the male explorer Amerigo Vespucci appearing fully dressed on the shore and a naked woman in a hammock serving as an image of the unknown space to conquer – while some cannibals are happily roasting a human leg in the background – it casts the 'discovery' of the New World as both a sexual encounter and a mythic moment of original inscription: the Florentine explorer first takes possession of the feminized land, then baptizes an entire continent which proceeds to enter history as the feminized version of its 'discoverer's' name.[19]

Gernon's anxiety about the lack of male authority repeats this theme: 'It is now since she [Ireland] was drawne out of the womb of rebellion about sixteen yeares . . . *and yet she wants a husband*, she is not embraced, she is not hedged and diched, there is no quicksett put into her.'[20] But while comparisons between the discursive appropriations of Ireland and the New World come readily to hand, no straightforward equations do justice to the specific parameters of their individual contexts. One immediate reminder of the different frameworks within which both discourses operate is Gernon's initial reference to a map of Europe. For him, Ireland figures not as part of a comparison with the 'alien cultures' of the New World but is firmly situated in a European arena. This is relevant in itself, if only to cast doubt on some recent critical views which argue for direct parallels between the English involvement in Ireland and the Americas.[21] Two further observations require comment here. The first is that the political implications emerging from a reading of feminized maps as allegories of power relations are difficult to control when the object of cartographic description is a European country, not a far-flung colony.[22] Europa, the daughter of Agenor, *is* female, of course, having been the rape victim of Zeus and only later elevated into the status of a continent. Yet Europe is clearly not open to metaphorical male penetration in the same way that a figure of sexual exploitation serves to define colonial conquest. This is rather a variation of the moment of aporia that English courtiers had to contend with in their rhetorical and military defence of a female monarch:[23] the need to demonstrate explicitly male sexual prowess while acting in the name of a queen who had chosen to raise virginity into a political symbol of national inviolability.

At the same time it is important to remember that England's overseas colonial activities – and this is the second point – had not yet moved in any significant sense beyond the stage of wishful thinking. Elizabethan England never properly owned an inch of land in the New World, and English colonial projects such as those propagated by the indefatigable Ralegh remained fantasies long into the seventeenth century. John Dee may well have been an ardent advocate of a 'British Empire', a phrase he was the first to use,[24] but the reality of England's New World presence still testified to the hollowness of that verbal construct. The same manifestly does not apply to Ireland. The English presence on Irish soil was a reality that stretched back to Henry II's twelfth-century campaign, chronicled by Giraldus Cambrensis, and the long sixteenth-century reconquest was pursued with renewed vigour in the light of Ireland's strategic position as a largely Catholic country in the immediate vicinity of Protestant England's continental (and similarly Catholic) rivals France and Spain.

Whatever Gernon's feminized Ireland was meant to signify, then, it is no simple equivalent of Galle's nude America reclining in a hammock, since the discursive construct of the 'virgin soil' that informed the land-as-female trope in writing on the New World could never apply in exactly the same sense to Ireland. Here, the narrative of English conquest is structured according to a logic related to, but nevertheless different from, the 'classic colonial triangle' consisting of 'pure' land, newly arrived colonizer and barbarous indigenous population, which Peter Hulme identifies as the central structural matrix of Galle's engraving.[25] For in Ireland, the long history of mutual contact did not make Irish space in any meaningful sense 'empty' or 'unknown' to English observers; rather, it remained riddled with ambiguity, being simultaneously a constituent part, and the conceptual opposite, of the national territory. As the analysis of its maps has shown, Ireland could only assume the degree of 'emptiness' necessary to enact the fiction of historical beginnings, and thus properly occupy its position in the colonial triangle, if it was rhetorically cleansed of historical meaning first; and as we have seen, this ideological move was fraught with considerable difficulties. This is not to say that the discursive separation of land and inhabitants in the New World did not prompt equally significant rhetorical efforts, but merely to insist on a relevant distinction emerging from the historical and geographical burden of Anglo-Irish proximity. It is here that the novel concept of cartographic sight could intervene and recast the narrative of Irish savagery as a self-inflicted apocalypse. In what remains of this final chapter, I will trace the articulation of this discursive logic in the employment of the trope of cannibalism in contemporary accounts of Ireland.

Irish cannibals are not an invention of the sixteenth century. Classical sources, frequently quoted even if rarely uncritically accepted as true,[26] tended to identify the Irish as anthropophagi – eaters of human flesh, known to live near Scythia – if they took note at all of the islanders at the very periphery of the ancient world:

> Concerning this island [Ireland] I have nothing certain to tell, except that its inhabitants are more savage than the Britons, since they are maneaters as well as heavy eaters, and since, further, they count it an honourable thing, when their fathers die to devour them . . . but I am saying this only with the understanding that I have no trustworthy witness for it.[27]

Strabo questions his sources here, but this did not prevent Irish cannibals from surviving well into the medieval period, as was evident from the more recent accounts of Irish anthropophagy available in *Holinshed's Chronicles*. John Hooker's translation of Giraldus Cambrensis' twelfth-century report on Ireland, for instance, contained the graphic description of a cannibalistic act reminiscent of a religious ritual. Dermon Mac Morogh, examining three hundred heads of slaughtered enemies, 'laid & put at his feet', discovered among these 'the head of one, whom especiallie and aboue all the rest he mortallie hated. And he taking vp that by the heare and eares, with his teeth most horriblie and cruellie bit awaie his nose and lips.'[28] Such examples of Irish anthropophagy, occurring at the 'frontiers of our' – and the Elizabethans' – 'time and space',[29] may be classified as illustrations of the wild and dark state of unbridled nature, a barbaric past of 'non-civilization', a historical time prior to the event defined as the arrival of culture. In other words, Irish cannibalism here emerges as a form of human wildness that existed before Christianization, before the English conquest, lending only further weight to the fiction of a 'civilizing mission'.

John Hooker's description of the Desmond Rebellion (1579–83), however, which was included in the second edition of *Holinshed's Chronicles* (1587) as a continuation of his translation of Giraldus, features Irish cannibals no longer safely located at the frontiers of time and space but living as contemporaries just across the Irish Sea. Facing the predicament of an 'extreme famine', the surviving victims of the rebellion were 'not onelie driuen to eat horsses, dogs and dead carions; *but also did deuoure the carcases of dead men*'.[30] Hooker's specific examples include the body of a 'malefactor' in Cork who – 'executed to death, and his bodie left vpon the gallows' – became a prey to 'certeine poore people [who] secretlie came, tooke him downe, and did eat him', and the drowned sailors of a ship 'lost through foule weather' who fell victim to the

> common people, who had a long time liued on limpets, orewads, and such shelfish as they could find, and which were now spent; as soone as they saw these dead bodies, they tooke them vp, and moste greedilie did eat and deuoure them: and not long after, death and famine did eat and consume them.[31]

Other observers also commented on this type of famine-induced survival cannibalism. Robert Payne accused the Earl of Desmond of having brought 'his countrie to that meserie that one did eate another for hunger',[32] and Edmund Campion, almost 20 years earlier, had reported on another famine in Ulster where '[the pore people] had first experienced many lamentable shiftes, as in scratching the dead bodyes oute of their graves, in whose sculles they boyled the same fleshe, and fedde thereof.'[33]

Yet, also on the pages of *Holinshed's Chronicles*, acts of pre-historic and survival cannibalism – explicable if horrific human actions – are registered alongside cannibalistic excesses implying quite a different motivation. During a rebellion in 1318, Hooker notes, '[i]t was said that the enimies . . . caused the lord Richards bodie to be cut in peeces, so to satisfie their malicious stomachs; but the same peeces were yet after-wards buried in the church of the friers minors at Limerike.'[34] Victims of cannibalism conventionally remain anonymous, and it may well be that 'lord Richards' established identity spared him, at least in the annals of Tudor historiography, from his terrible fate. Yet even if the rumours turned out to be false, the passage implies that they were not entirely unfounded. The casual reference to Shane O'Neill as 'that canyball'[35] in one of Sidney's letters to Leicester (1566) may be little more than an insult born out of frustration, but elsewhere references to the consumption of human flesh by the native Irish point to reasons of greater cultural urgency behind this imagery. Fynes Moryson gave this account of events in Ulster in 1602:

> Captaine Trevor & many honest Gentleman lying in the Newry can witnes, that some old women of those parts, used to make a fier in the fields, & divers little children driving out the cattel in the cold mornings, and comming thither to warme them, *were by them sur-prised, killed and eaten*, which at last was discovered by a great girle breakinge from them by strength of her body, and Captain Trevor sending out souldiers to know the truth, they found the childrens skulles and bones, and apprehended the old women who were executed for the fact.[36]

This scene is strangely mirrored by the report of William Farmer who in 1615 reversed the confrontation between the generations:

> As [Sir Arthur Chichester] travelled through a wood there was felt a great savour, as it were roasting or broiling of flesh; the governor sent out soldiers to search the wood, and they found a cabin where a woman was dead, and five children by her made fire to her thighs and arms and sides, *roasting her flesh and eating it*.[37]

A textual analysis of the last two passages needs to account for the suspiciously indirect routes through which knowledge of cannibalism reaches both narrators: Moryson points not to the observation of the act as such but to the remains of the cannibal banquet;[38] Farmer's report is twice removed from its original source, passing first from eyewitness to Chichester, and only from there to the narrator. A larger pattern may be discerned here, for both the metonymic identification through suspicious traces and the process of narrative transmission involving a series of previous accounts are rhetorical devices that have accompanied the European discourse on New World cannibalism from its inception in Columbus' 1493 *Journal*. It was there that the Genoese seafarer reported that the 'friendly' Arawaks had told him (in a language he had been exposed to for just six weeks) about the 'savage' people of the Caribs – supposedly eaters of human flesh – living on the next island.[39] In his controversial study of the 'man-eating myth', the anthropologist William Arens has detected these precise patterns in every single instance of reported cannibalism throughout human history. According to Arens, cannibalism cannot be proven to have ever existed in any part of the world, leading him to argue that cannibalism as a collective ethnographic myth is granted the force of fact without any need for verification either through scientific proof or eyewitness report.[40] But if the historical existence of regular anthropophagy as a dietary habit remains doubtful, its unparalleled usefulness as a discursive mechanism of exclusion does not: '[F]or the most part cannibalism is a discourse on the Other, defining out-groups in terms of their horrifying man-eating propensities.'[41]

If it is correct to say, then, that 'the idea of cannibalism', as Arens claims, is a cultural construction that always 'exists prior to and thus independent of the evidence',[42] how does one account for the frequency, across historical time, of cannibalistic imagery in Tudor and Stuart descriptions of Ireland? One reason is certainly the existence of a conventional set of ethnographic markers which had transformed the former *isla sanctorum* into an *isla barbarorum*, a major semantic shift Denis Bethell has identified as a result of the twelfth-century English conquest: '"Barbarity" had become, and was to remain, a cliché in describing the Irish'.[43] Thus, as Quinn has noted, in Elizabethan England '[t]he use of the Irish as the standard of savage and outlandish reference was well established by 1560'.[44] Indeed, as a descriptive formula enabling cultural comparisons, rhetorically invoked cannibals are a frequent occurrence in Irish tracts. Fynes Moryson thought that 'these wild Irish are not much unlike to wild beasts, in whose caves a beast passing that way, might perhaps find meate, but not without danger to be ill entertained, perhaps devoured of his insatiable Host';[45] Sir John Davies reckoned that the Irish 'were little better then *Canniballes*, who doe hunt one another; and hee that hath most strength and swiftnes, doth eate and deuoure all his fellowes';[46] and Barnabe Rich believed that '[t]he *Canibals*, deuourers of mens flesh, doe leaue to bee fierce amongst themselues, but the Irish, without all respect, are euer most cruel to their very next neighbours.'[47] The indiscriminate use of the trope of cannibalism in these quotations may be taken to confirm

Arens' recent observation that '[t]he theme of the "other" as cannibal had, and has, an existence beyond the control of those employing it at any particular moment.'[48]

Yet to insist merely on the persuasive force of time-honoured descriptive patterns does not go a long way to explaining the persistence with which 'real' Irish cannibals roam the textual landscapes of the English imagination. A more substantial explanation could be gained from the application to an Irish context of Hulme's analysis of the man-eating myth in the colonial encounter. In this reading, the identification of cannibalism as ' "the image of ferocious consumption of human flesh" frequently used to mark the boundary between one community and its others'[49] emphatically points to Ireland's function as an identity-forming device in the emerging conception of English nationhood: the cultural projection of Irish anthropophagy allows the drawing of borders, the setting up of a clearly discernible discursive terrain which defines acceptable versions of the civilized self precisely by describing its worst possible opposite, its antagonistic 'other', in order to articulate what this self is not, and must never degenerate into. The specific value of the cannibalistic fantasy would thus be its service to a political concept of collective identity defined against images of a depraved and threatening otherness, allowing the ferocious assault on the human body to be read simultaneously as a collective and an individual myth: in violating corporal boundaries, cannibals also violate the borders of the state, the body politic, as well as the divine order of earthly affairs, the macrocosm of creation, reflected in the microcosm of each individual human frame.

While the cultural significance of cannibalistic imagery may well be implicated in the early modern formation of national identity, its function in the Irish context principally reveals, in Hulme's formulation, how 'the topic of land is dissimulated by the topic of savagery'.[50] The motif of self-destruction inherent in the trope of Irish cannibalism – at work, for instance, in Hooker's account of the Desmond Rebellion where 'death and famine did eat and consume' the savages 'not long after' their horrific deed – points to its usefulness as a rhetorical cleansing device which served to erase the disruptive presence of the native Irish from a landscape that was redefined both as a profitable arena for English settlers and as part of the national territory – a concept promoted, for instance, by Davies' imperial fantasy of an 'English race and Nation'[51] fully incorporating Ireland.

The version of Ireland as a promising site for commercial ventures was readily available in English accounts. One late example by Thomas

Gainsford is typical in that it proceeds from a description of place to a description of people, but only after an initial reminder of the local dangers associated with the

> vnfirmenes of ground, & the lurking rebell, who will plash downe whole trees ouer the paces and so intricately winde them, or lay them, that they shall be a strong barracado, and then lurke in ambush amongst the standing wood, playing vpon all commers as they intend to goe along.[52]

However, once the English presence in Ireland has taken root, the text suggests, all this will no longer be a problem. Running through each of the four historical provinces individually, Gainsford proleptically offers up Ireland as a peaceful place of unlimited natural abundance: Leinster, 'more orderly than the rest', is 'very fruitfull and pleasant'; the Munster coastline can boast 'many excellent harbours' whose adjacent grounds 'are very fertile'; the fish caught in Connaught – 'Salmon, Breame, Pike, and diuers other sorts' – are so huge that Gainsford fears he may not be believed if he gave the true figures; and the lakes in Ulster are known to yield trout an incredible '46 inches long'.[53]

This image of Ireland as a shamefully neglected space of unused resources and cornucopian splendour was frequently the object of English praise:

> [Ireland] is a moste bewtifull and swete Countrie as anye is vnder heaven, seamed thoroughe out with manye goodlye rivers replenished with all sortes of fishe moste aboundantlye sprinckled with manye swete Ilandes and goodlye lakes like little Inlande seas, that will carye even shipps vppon theire waters, adorned with goodly woodes fitt for buildinge of howsses and shipps so comodiously as that if some princes in the worlde had them they they [*sic*] woulde sone hope to be Lordes of all the seas and ere longe of all the worlde.[54]

Almost two decades after Spenser offered this eulogy to the readers of his *View of the Present State of Ireland* (1596), Davies endorsed an equally Edenic view of Ireland as a land of milk and honey, or rather as a space of unparalleled natural advantages and willing, serviceable bodies, all ready for economic exploitation:

> I haue obserued the good Temperature of the Ayre; the Fruitfulnesse of the Soyle; the pleasant and commodious seats for habitation; the

safe and large Ports and Hauens lying open for Trafficke, into all the West parts of the world; the long Inlets of many Nauigable Riuers, and so many great Lakes, and fresh Ponds within the Land; (as the like are not to be seene in any part of Europe) the rich Fishings, and Wild Fowle of all kinds; and lastly, the Bodies and Minds of the people, endued with extraordinarie abilities of Nature.[55]

The principal reason for the inclusion of such passages in tracts that elsewhere paraded images of destruction and savagery is the attempt to make Ireland an attractive option for new English (and Scottish) settlers who needed to be recruited in high numbers if the Elizabethan reconquest and the plantation projects in its wake were to have any chance of lasting success.

In descriptions of Ireland such as these the precepts of a Christian 'civilizing mission' – which could boast success once cannibalism had ceased – is often much less noticeable than a rampant land-grabbing impulse (which could be read, I suppose, as an ironic reversal of the accusation of cannibalism levelled at the Irish). Their rhetorical function is much more complex than their historical role as colonial advertisements reveals. The dream of Biblical splendour reappears as the opening image of an infamous passage in Spenser's *View* which describes in graphic detail the dreadful consequences of a famine in Munster. This account, which has always been among the *View*'s most frequently quoted passages, has recently attracted much politically committed commentary. In joining this chorus of critical voices my intention is not to repeat what has been said before but to draw attention to what I suggest is the inherent cartographic rationale of Spenser's disturbing vision. '[I]n Those late warrs of mounster,' he writes, again with reference to the Desmond Rebellion,

for notwithstandinge that the same was a most ritche and plentifull Countrye full of Corne and Cattell that ye would haue thoughte they Coulde haue bene able to stande longe yeat ere one yeare and a haulfe they weare broughte to soe wonderfull wretchednes as that anie stonie harte would heue rewed the same. Out of euerie Corner of the woods and glinnes they Came Crepinge forthe vppon theire handes for theire Leggs Coulde not beare them, they loked like Anotomies of deathe, they spake like ghostes Cryinge out of their graues, they did eate the dead Carrions, happie wheare they Coulde finde them, Yea and one another sone after, in so muche as the verye carkasses they spared not to scrape out of their graues.[56]

The crucial question to ask about this description is not, of course, whether Spenser was really an eyewitness (as he claims),[57] but what descriptive strategies he employs and to what purpose. The Irish are first divested of their human attributes: unable either to walk upright or to speak properly,[58] they emerge from the woods and glens like beasts, 'Crepinge forthe vppon theire handes'. Their appearance is couched in images of death, anticipating their imminent fate: they are 'Anotomies of deathe', 'ghostes Cryinge out of their graues', they devour 'carkasses' and 'dead Carrions'. Mediating between life and death, courting famine, cannibalism and physical self-destruction, the native Irish implicated in this scene of horror suffer the inevitable consequence of what the *View* as a whole contends to be the subhuman condition of barbarism almost inextricably ingrained in the bodies of the people.

The main point I wish to make about this passage concerns its spatial framework, a subtle progression of three distinct points of view following each other in quick succession. Just prior to the emergence of the Irish from the hidden corners of the woods and glens, the narrator focuses on Munster's Arcadian landscape, 'a most ritche and plentifull Countrye full of Corne and Cattell'. Next, this totalizing cartographic view, which registers the land as a pastoral idyll, is interrupted by what purports to be an eyewitness report woven into the textual fabric. Instead of continuing to gaze at the wide and undifferentiated expanse we are now confronted with single figures and individual landscape elements, 'anotomies', 'ghostes', 'graves'. Thus, just as our view zooms in on a distinct selection of people and spatial markers, an initially peaceful landscape transforms into an image of death and decay. Put differently, removed from the lofty position of the map we had been contemplating we find ourselves suddenly down on ground level, wandering among the starving creatures driven to the extreme measure of cannibalism. This scalar descent is reversed in the immediate continuation of the passage:

> [I]n shorte space theare weare non allmoste lefte and a moste populous and plentifull Countrye sodenlye left voide of man or beaste, yeat sure in all that warr theare perished not manie by the sworde but all by the extreamitye of famine which they themselves had wroughte.[59]

Spenser's rhetorical wheel has come full circle and we are back where we started, gazing at a map of the entire territory, 'a moste populous and plentifull Countrye'. The representational surplus gained by this sleight of hand is immediately apparent, for what was before a rich and fertile if sadly mistreated landscape, unfortunately populated by a bunch of people ready to stage a rebellion to enforce the priority of their claim to ownership, is now transformed into the colonial fantasy of a spatial *tabula rasa*, a land 'sodenlye left voide of man or beaste'.

The point of this reading is twofold. First, it aims to highlight the function in an Irish context of the distinction between itinerary and map I have identified as the central representational conflict with regard to the narrative construction of national space in Britain. Spenser's description oscillates back and forth between both paradigms, replacing the initial view of global fixity provided by the map with the close-up vision of an itinerary unable to lock its object of description – the starving population – in the secure still life of a cartographic frame. The link between both paradigms is imagined as a hermeneutic sequence. While the close inspection enabled by the itinerary is an almost natural stop on the way to a fuller comprehension of the spatial stories enacted below, it is the shocking realization of the horrid details that allows the righteous acceptance of the people's erasure. At the same time the moments of visual distance opening and ending the passage are not simply alternatives to the immediacy of contact but a higher order of spatial control, and the totalizing temporality of the landscape produced by the cannibalistic assault – an apocalypse for the Irish, an act of creation for the English – thus marks the 'just' transfer of property from a state of savagery to a state of civilization. But the logic behind this usage of space as a defining category of the people it signifies, implies a relationship of structural similarity rather than of utter contrast: in staging this mechanism of exclusion it re-enacts for Ireland what was observable in England, the success of the cartographic over the performative paradigm in the narration of the nation.

Second, my reading is intended to emphasize the instrumentalization of cartography as an effective means of labelling the Irish as hopelessly self-destructive savages. The indebtedness of the passage to powerful tropes of colonial discourse is self-evident. The circular movement of Spenser's rhetoric clearly serves to transform well-populated Ireland into the conspicuously 'uninhabited isle' of *The Tempest*. The metonymic progress from fragment to whole suggests that in the whole of Ireland there might equally be 'non allmoste lefte'; the passage thus employs in its representational apparatus a figure that Greenblatt has defined as one of the characteristic tropes of colonial appropriation: 'The supplement that imagination brings to vision expands the perceptual field, encompassing the distant hills and valleys or the whole of an island or an entire continent, and the bit that has actually been seen becomes by metonymy a representation of the whole.'[60] In Spenser's *View*, the final comment lays all the blame on the victims themselves: nothing has happened, no cannibalistic excess, no English scorched-earth policy, no acts of utter desperation, but what 'they themselves had wroughte.' The narrative of cannibalism, and the tale of Irish savagery more at large, can thus be grasped as a spatially generated discourse that offers to do in rhetoric what the use of geometry in political reform plans or the visual play of absence and presence on presentational maps could each achieve in their respective epistemological modes: to write the 'groundless' fiction of a fully anglicized extension of the national sphere, encompassing Irish space and Irish bodies alike. The analysis of spatial *narratives* thus confirms what I have argued throughout this study, that the singular achievement of a new sense of space fostered by modern cartography, be it in England, Ireland or elsewhere, is little understood if in our continuing deference to the visual fantasy of unlimited telescopic sight we fail to take note of its own darker purpose.

Notes

Introduction: the Cartographic Transaction

1. In his commentary on the painting in *Vermeer. Das Gesamtwerk* (Stuttgart/Zurich: Belser, 1995), 170, Arthur K. Wheelock, Jr, argues that the geographer gazes thoughtfully through the window; the curators of the Städel, according to an explanatory note put up next to the painting in 1999, think that his 'expression and bodily posture indicate the moment of recognition after a concentrated phase of contemplation'. The geographer may indeed be lost in thought and not looking at anything in particular but the suggestion of a sudden moment of sublime recognition seems to me entirely conjectural. I find no support for it in the painting.

2. This is confirmed by recent examinations of the painting with infrared reflectography which suggest that Vermeer may have seen the need to change the position of the geographer's head. At an earlier stage in the composition of the painting, the geographer appears to have been looking downward at the map on the table before him. Cf. Peter Waldeis, 'Erste gemeinsame Untersuchung des *Geographen* und *Astronomen*', *Johann Vermeer. Der Geograph und der Astronom* (Frankfurt: Städelsches Kunstinstitut, 1997), 39–46, 44. If this assumption is correct, it is surely significant that Vermeer, in revising the theme of his painting, should have consciously introduced into the representational scene of the canvas a moment of visual contact between the geographer and the window.

3. I should also draw attention to the upper section of the window which is standing conspicuously open. Such open windows (or doors), through which an outside world infiltrates the 'domestic haven', are features common to Vermeer's interiors.

4. And note how three of these objects – the globe, the map on the table and that on the floor – emerge from the shadows to be caught in the light coming through the window, a pattern that establishes a clear link between these items of geographical interest, the window and the geographer's face.

5. Radolph Agas, *A Preparative to Platting of Landes and Tenements for Surueigh* (London: Thomas Scarlet, 1596), 4–5 (my italics).

6. Richard Helgerson, 'Genre Painting, Maps, and National Insecurity in Seventeenth-Century Holland', MS 3. A German translation of this essay appeared in Ulrich Bielefeld and Gisela Engel (eds), *Bilder der Nation. Kulturelle und politische Konstruktionen des Nationalen am Beginn der europäischen Moderne* (Hamburg: Verlag Edition, 1998), 123–52. I am grateful to Richard Helgerson for providing me with a typescript of the English version of his essay in advance of publication. For a general analysis of the significance of maps in seventeenth-century Dutch genre painting see Bärbel Hedinger, *Karten in Bildern. Zur Ikonographie der Wandkarte in holländischen Interieurgemälden des 17. Jahrhunderts* (Hildesheim et al.: Solms, 1986).

7. The perspectival composition of the painting is analysed in detail by Jørgen Wadum, 'Vermeers zielgerichtete Beobachtungen', *Johann Vermeer. Der Geograph und der Astronom*, 31–8.

8. David Harvey, *The Condition of Postmodernity. An Enquiry into the Origins of Cultural Change* (Cambridge, MA: Blackwell, 1989), 244. Harvey is referring to an argument put forward by Samuel Edgerton who identifies the Ptolemaic theory of map projection as the central impulse for the Renaissance rediscovery of linear perspective. See 'From Mental Matrix to *Mappamundi* to Christian Empire: the Heritage of Ptolemaic Cartography in the Renaissance', David Woodward (ed.), *Art and Cartography. Six Historical Essays* (Chicago: Chicago University Press, 1987), 10–49. Edgerton has been taken to task by Svetlana Alpers who sees no 'determining link between Albertian perspective and Ptolemaic recipes for map projection' (70), and argues instead for an essential difference between maps and perspective paintings in terms of their pictorial conception. See 'The Mapping Impulse in Dutch Art', in Woodward (ed.), *Art and Cartography*, 51–96.

9. Michel Foucault, 'Of Other Spaces' [1969], *Diacritics*, 16 (1986) 22–7, 22.

10. Frank Lestringant, *Mapping the Renaissance World. The Geographical Imagination in the Age of Discovery* (Cambridge: Polity Press, 1994 [French original 1991]), 1.

11. Ibid., 4.

12. John Hale, *The Civilization of Europe in the Renaissance* (London: Harper-Collins, 1993), 15.

13. For an account of the early modern period as an epoch fuelled by a secular reappreciation of the material object, see Lisa Jardine, *Worldly Goods* (London: Macmillan, 1996).

14. Hale, *Civilization of Europe*, 16.

15. The two geographers who revolutionized their field – Gerard Mercator and Abraham Ortelius – represent the opposite poles of this spectrum: for the older Mercator cartography was a scholarly pursuit, and the successive states of his maps are evidence of a continuous process of critical reassessement; for Ortelius, who emerged from the mercantile atmosphere of sixteenth-century Antwerp, maps meant business and new or updated geographic information was primarily a financial asset. See C. Koeman, *The History of Abraham Ortelius and His Theatrum Orbis Terrarum* (Lausanne: Sequoia SA, 1964), 20.

16. Denis Cosgrove, 'Introduction: Mapping Meaning', in Cosgrove (ed.), *Mappings* (London: Reaktion, 1999), 1–23, 9.

17. Lestringant, *Mapping the Renaissance World*, 7.

18. For a seminal analysis of the national dimension of early modern English chorography see Richard Helgerson, *Forms of Nationhood. The Elizabethan Writing of England* (Chicago: Chicago University Press, 1992), chapter 3: 'The Land Speaks', 105–47.

19. See Henri Lefebvre, *The Production of Space*, trans. Donald Nicolson-Smith (Oxford: Blackwell, 1991 [French original 1974]).

20. Ibid., 27.

21. Edward W. Soja, *Postmodern Geographies. The Reassertion of Space in Critical Social Theory* (London: Verso, 1989), 7.

22. J. B. Harley and David Woodward, 'Preface', in Harley and Woodward (eds), *History of Cartography, Vol. 1: Cartography in Prehistoric, Ancient, and Medieval Europe and the Mediterranean* (Chicago: Chicago University Press, 1987), xvi.
23. John Gillies, *Shakespeare and the Geography of Difference* (Cambridge: Cambridge University Press, 1994), 54.
24. The phrase is Denis Cosgrove's. See his 'Mapping Meaning', 1.

Part I Measurements

1. As the prologue of *Tamburlaine the Great*, Part 1, announces. I quote throughout from *Christopher Marlowe. The Complete Plays*, ed. J. B. Steane (Harmondsworth: Penguin, 1969).
2. Thomas Hood, *A Copie of the Speech Made by the Mathematical Lecturer* [delivered 4 November 1588], (London: Edward Allde, 1588), sig. A2v. *Tamburlaine* is usually dated to 1587/88.
3. The relevant discussions of the play's cartographic background are Ethel Seaton, 'Marlowe's Map' [1924], Clifford Leech (ed.), *Marlowe: A Collection of Critical Essays*, Twentieth-Century Views (Englewood Cliffs, NJ: Prentice Hall, 1964), 36–56; and John Gillies, 'Marlowe, the *Timur* Myth, and the Motives of Geography', John Gillies and Virginia Mason Vaughan (eds), *Playing the Globe: Genre and Geography in English Renaissance Drama* (Madison, WI: Fairleigh Dickinson University Press, 1998). I am grateful to John Gillies for providing me with a typescript of his essay in advance of publication. For a discussion exploring links between Marlowe's use of cartography and contemporary theories of warfare see Nick de Somogyi, 'Marlowe's Maps of War', Darryll Grantley and Peter Roberts (eds), *Christopher Marlowe and English Renaissance Culture* (Aldershot: Scolar Press, 1996), 96–109.
4. Seaton traces Techelles' exact route through Africa ('Marlowe's Map', 40–2), calling the report 'a passage in which one can almost follow Marlowe's finger travelling down the page as he plans the campaign' (42).
5. Such a usage of his atlas is ultimately suggested by Ortelius himself: in the preface he explains the geographical arrangement of the maps in the collection by taking the reader on a tour around his own world map.
6. See also Chapter 4 below.
7. Quoted from the only contemporary English edition: Abraham Ortelius, *The Theatre of the Whole World* (London: Iohn Norton and Iohn Bill, 1606), preface (my italics).
8. John Gillies, *Shakespeare and the Geography of Difference* (Cambridge: Cambridge University Press, 1994), 57.
9. Ibid.

Chapter 1 Mathematics of the World

1. '[I]l ne fût accusé d'aucun autre crime, que d'avoir eu dans son cabinet une carte du globe terrestre'. *Histoire Romaine e'crite par Xiphilin, par Zonare, et par Zosime*, trans. L. Cousin (Paris: P. Rocolet, 1678), 332 (my translation).

2. In German usage, where the book is usually referred to as *Die Schedelsche Weltchronik*, author takes priority over place. The book is available in several modern editions.

3. Adrian Wilson, *The Nuremberg Chronicle Designs* (San Francisco: Roxburghe Club, 1969), n.p.

4. A useful and informative study of the material context of the *Nuremberg Chronicle* is Elisabeth Rücker, *Hartmann Schedels Weltchronik. Das größte Buchunternehmen der Dürer-Zeit* (Munich: Prestel, 1988). Cf. also Stephan Füssel, 'Die Weltchronik – eine Nürnberger Gemeinschaftsleistung', *500 Jahre Schedelsche Weltchronik*, Pirckheimer-Jahrbuch 9, ed. Stephan Füssel (Nuremberg: Carl, 1994), 7–30.

5. Cf. Rücker, *Hartmann Schedels Weltchronik*, 84–9.

6. Rücker includes an appendix which divides Schedel's cities into 'authentic' and 'imaginary' views.

7. See Max Schefold's preface to Georg Braun and Frans Hogenberg, *Beschreibung und Contrafactur der vornembster Stät der Welt* [1572–1617], colour facsimile edition (Stuttgart: Müller & Schindler, 1965), 5–6.

8. By Johann Schönsperger in 1496. This affordable and cheaply produced 'people's version' ran to several editions. Cf. Füssel, 'Die Weltchronik', 25.

9. Sebastian Brant, *The Shyp of folys of the worlde*, trans. Alexander Barclay (London: Rycharde Pynson, 1509), fol. 138v (wrongly paginated 140v)–139r.

10. Brant, *Shyp*, fol. 140r. Barclay generally takes great liberties with Brant's poem but this passage is fairly close to the original: 'So ist doch das eyn grosser tor/Der in seym synn wygt so gering/Das er well wissen frembde ding/Und die erkennen eygentlich/Und kan doch nit erkennen sich'. Sebastian Brant, *Das narren schyeff* (Nuremberg: Peter Wagner, 1494), sig. N1v.

11. Brant, *Shyp*, fol. 140r.

12. See Numa Broc, *La géographie de la Renaissance, 1420–1620* (Paris: CHTS, 1986 [First published 1980]), 67.

13. See Caterina Albano, 'Visible Bodies: Cartography and Anatomy', in Andrew Gordon and Bernhard Klein (eds), *Literature, Mapping, and the Politics of Space in Early Modern Britain* (Cambridge University Press, forthcoming); and Hannes Kästner, 'Der Arzt und die Kosmographie', in Ludger Grenzmann and Karl Stackmann (eds), *Literatur und Laienbildung im Spätmittelalter und in der Reformationszeit* (Stuttgart: Metzler, 1984), 504–31.

14. The figures are taken from Wilson, *The Nuremberg Chronicle Designs*, preface, who further points out that the book contains 'some 1800 illustrations provided by the multiple use, for different subjects, of 645 beautiful woodblocks' (n.p.).

15. Brant, *Shyp*, fol. 139r (my italics).

16. See Frank Lestringant, *Mapping the Renaissance World. The Geographical Imagination in the Age of Discovery* (London: Polity Press, 1994 [French original 1991]), 1–11.

17. Petrus Apianus, *Cosmographicus liber* (Landshut: by the author, 1524). Originally published in Latin, the book remained in print in various languages throughout the century (no English translation, either contemporary or

modern, exists). The illustrations (Figure 1.3), unaltered in substance from the first edition, are taken from the 1550 Latin edition published in Antwerp.

18. The analogy, widespread in the sixteenth century, might result from a misreading of Ptolemy who established no necessary connection between geography and painting. See Lucia Nuti, 'Mapping Places: Chorography and Vision in the Renaissance', in Denis Cosgrove (ed.), *Mappings* (London: Reaktion, 1999), 90–108, 90–1.

19. Except for twentieth-century astronauts, I should perhaps add, but even for modern technical man the view from outer space has been a profoundly disconcerting experience.

20. Anthony Grafton, *New Worlds, Ancient Texts. The Power of Tradition and the Shock of Discovery* (Cambridge, MA/London: Belknap Press at Harvard University Press, 1992), 13.

21. Though one of the foremost German humanist scholars, Konrad Celtis, was meant to update the chronicle for a future edition which never materialized.

22. Sebastian Münster, *Cosmographei oder beschreibung aller länder, herschafften, fürnemsten stetten, geschichten, gebreüchen, hantierungen etc.* (Basle: Heinrich Petri, 1544). Editions of the *Cosmographei* were continuously enlarged and frequently reprinted throughout the sixteenth and early seventeenth centuries, with the 1550 and 1588 editions containing the most significant alterations from a cartographic point of view (see also further below). I will be quoting from the 1592 edition throughout. A facsimile of the 1628 edition has been published in Lindau (Antiqua, 1984). For a comprehensive list of contemporary editions see Karl Heinz Burmeister, *Sebastian Münster. Eine Bibliographie* (Wiesbaden: Guido Pressler, 1964).

23. See Broc, *La géographie de la Renaissance*, chapters 5 and 6.

24. Georg Braun and Frans Hogenberg, *Civitates Orbis Terrarum*, 6 vols (Antwerp: Filips Galle/Cologne: apud Auctores, 1572–1617), vol. 1, preface. A facsimile reproduction has been published in Cleveland and New York (World Publishing, 1966).

25. Though the frequent inclusion of city views in atlases and wall maps from the early seventeenth century onwards may of course be seen as a response to precisely this marginalization of the city in the actual map image itself.

26. Prominent among these are Guillaume Guerroult, *Epitome de la Corographie d'Europe, illustré des pourtraitz des Villes plus renommees d'icelle* (Lyons: Balthazar Arnoullet, 1553); Antoine du Pinet, *Plantz, povrtraitz et descriptions de plvsievrs villes et forteresses* (Lyons: Iean d'Ogerolles, 1564); Philipp Apian, *Baierische Landtaflen* (Ingolstadt: n.p., 1567); Ludovico Guicciardini, *Descrittione di tvtti i paesi bassi* (Antwerp: Guglielmo Siluio, 1567); M. Guilio Ballino, *De disegni delle piu illustri citta, & fortezze del mondo* (Venice: Bolognino Zaltieri, 1569); Francesco Valegio and Martin Rosa, *Raccolta di le piu illustri et famose citta di tvtto il mondo* (Venice: n.p., 1579); Abraham Saur, *Parvum Theatrum Urbium* (Frankfurt: Nicolaus Bassäum, 1581).

27. 'Wie ... kunstreiche Bawmeister *den gantzen erdenkreiß* ... mit Stäten und flecken verziert, haben mit höhester scharpfsinnigkeyt und wunderbar-

lichem fleiß Simon Nouellanus und Franß Hogenberg so artlich, lebendig
vnd mit aller Städte proportion, gelecht, vnd gestalt an tag gethan, das man
nit deren ebenbildt vnd contrafactur: sonder die stet selbs fur den augen
scheint zu haben.' Georg Braun and Frans Hogenberg, *Beschreibung und Contrafactur der vornembster Stätt der Welt* (Cologne: Heinrich von Ach, 1574),
preface, n.p. (italics and translation mine).

28. All in volume 4, except Kostajnica (volume 6).

29. I am referring to the *Index Alphabeticus prior singularum regionum* added to
    volume 6 (published in 1617) which specifies that 'Ciuitates vel oppida in
    plano, vel prospectiue, vt vocant, effigiatas esse denotant.' Of those marked,
    276 images are in the category 'prosp', 281 in the category 'pl'.

30. See Andrew Gordon, 'Performing London: The Map and the City in Cere-
    mony', Gordon and Klein (eds), *The Politics of Space* (forthcoming).

31. Abraham Ortelius, *Theatrum Orbis Terrarum* (Antwerp: A. C. Diesth, 1570).
    A black and white facsimile of the first edition has been published by Merid-
    ian (Amsterdam, 1964) with an introduction by R. A. Skelton. A colour
    facsimile of the first edition was issued in the same year in Lausanne, accom-
    panied by the separate publication of an account of Ortelius' life and his
    Antwerp workshop: C. Koeman, *The History of Abraham Ortelius and His The-
    atrum Orbis Terrarum* (Lausanne: Sequoia SA, 1964).

32. Leo Bagrow, *History of Cartography*, second edition, revised and enlarged by
    R. A. Skelton (Chicago: Precedent Publishing, 1985), 179.

33. Gerard Mercator, *Atlas sive Cosmographicae Meditationes de fabrica mundi et
    fabricati figura* (Düsseldorf: Albertus Busius, 1595). Individual sections of this
    atlas had been appearing separately since 1585. Mercator died in 1595 and
    did not live to see the the complete edition published.

34. See Koeman, *The History of Abraham Ortelius*, for a list of editions and their
    variants.

35. John Gillies, *Shakespeare and the Geography of Difference* (Cambridge:
    Cambridge University Press, 1994), 62.

36. Ironically, the illustration is very close to the now discarded, old scale-map
    of Europe, included in earlier editions of the *Cosmography*. Its more imme-
    diate source, however, is the woodcut 'Europa prima pars terrae in forma
    virginis', included in Heinrich Bünting, *Itinerarium totius sacrae scripturae*
    (Wittenberg: Zacharias Krafft, 1587) (earlier editions of Bünting do not
    contain the map). This map was almost certainly intended 'to personify
    Spanish domination and tyranny of Europe which reached its apex under
    the reign of Philip II' (*The Discovery of the World. Maps of the Earth and the
    Cosmos from the David M. Stewart Collection* (David M. Stewart Museum, St
    Helen's Island, Montreal: University of Chicago Press, 1985), 81). Although
    the Spanish imperial crown is retained on the map in Münster, its new
    context, specifically the passage on Europe's 'fertility' on the page facing it
    in the *Cosmography*, substantially alters the significance of the image, as I
    suggest below. The earliest 'virgo' map of Europe is a woodcut by Johann
    Putsch (Paris: Christian Wechel, 1537), intended to glorify the Habsburg
    rule over Europe. It is again revived later in the century in the anti-Spanish
    political pamphlet *Het Spaens Europa* (Amsterdam: n.p., 1598). For detailed
    historical information on these maps see H. A. M. van der Heijden, *De oudste
    gedrukte Kaarten van Europa* (Alphen aan den Rijn: Canaletto, 1992). For a

feminist reading of Bünting's map as a semiological system see Annegret Pelz, *Reisen durch die eigene Fremde. Reiseliteratur von Frauen als autogeographische Schriften* (Cologne et al.: Böhlau, 1993), 13–45.

37. For a general account of the gendered sense of space in the early modern period see Sabine Schülting, *Wilde Frauen, fremde Welten. Kolonisierungsgeschichten aus Amerika* (Reinbek: Rowohlt, 1997), 24–46.

38. 'Gleicherweiß wie der Himmel Gottes Wohnung ist, also ist das Erdtrich der Menschen und Thieren Behausung, *ja ihre Mutter*. Dann es empfaht uns so wir geboren werden, es ernehrt und tregt uns dieweil wir leben, und zu letzt empfaht es uns in sein Schoß, behelt unser Cörper biß zum Jüngsten tag'. Münster, *Cosmographei*, ed. 1592, 4 (my italics and translation).

39. Ibid., 4–5.

40. Ibid., 5.

41. Viktor Hantzsch, in *Sebastian Münster. Leben, Werk, wissenschaftliche Bedeutung*, Abhandlungen der philologisch-historischen Classe der Königlich-Sächsischen Gesellschaft der Wissenschaften 18 (Leipzig: B. G. Teubner, 1898), 67, suggests 1650 as the year of the last edition but Burmeister, *Sebastian Münster*, finds no trace of an edition after 1628.

42. Ortelius, *The Theatre of the Whole World*, preface (n.p.).

43. John Speed, *The Theatre of the Empire of Great Britaine* (London: Iohn Sudbury & George Humble, 1611), 1 (Speed's italics).

44. See his *The Body Emblazoned. Dissection and the Human Body in Renaissance Culture* (London: Routledge, 1995), 3 (*et passim*).

45. Gillies, *Geography of Difference*, 188.

Chapter 2 Land Measuring: an Upstart Art

1. William Cuningham, *The Cosmographical Glasse* (London: John Day, 1559).

2. Richard Eden published the sections on the New World as *A Treatyse of the Newe India* (London: Edward Sutton, 1553), and later added *A Briefe Collection and compendious extract of straunge and memorable thinges, gathered out of the Cosmographye of Sebastian Munster* (London: Thomas Marshe, 1572; reprinted 1574). George North translated the material on Scandinavia in 1561. Eden's *The History of Travayle in the West and the East Indies* (London: Richard Jugge, 1577), contains selections from Münster alongside other material, principally from Peter Martyr's *Decades*. Karl Heinz Burmeister, *Sebastian Münster. Eine Bibliographie* (Wiesbaden: Guido Pressler, 1964), entry 106, lists another contemporary edition of which no trace survives: *An abridgment of S. Munster's Chronicle* (London: W. Marshal, 1552).

3. Gerard Mercator (with Jodocus Hondius), *Atlas, or, a Geographicke description of the Regions, Countries and Kingdomes of the world*, trans. H. Hexham, 2 vols (Amsterdam: n.p., 1636–8). John Speed's *Prospect of the Most Famous Parts of the World*, published in 1627, was a shorter English version of a world atlas.

4. Thomas Hood, *A Copie of the Speech Made by the Mathematical Lecturer* (London: Edward Allde, 1588), sig. B3r.

5. Robert Recorde, *The ground of artes* (London: Reynold Wolfe, 1543), 1v.

6. Robert Recorde, *The pathway to knowledg* [*sic*] (London: Reynold Wolfe, 1551), preface.
7. Ibid.
8. John Fitzherbert, *The boke of surueying and improumentes* (London: Rycharde Pynson, 1523), sig. H1r.
9. Andrew McRae, *God Speed the Plough. The Representation of Agrarian England, 1500–1660* (Cambridge: Cambridge University Press, 1996), chapter 6: 'To Know One's Own: the Discourse of the Estate Surveyor', 169–97, 172.
10. See the chapter 'Surveying in Medieval England' in R. A. Skelton and P. D. A. Harvey (eds), *Local Maps and Plans from Medieval England* (Oxford: Clarendon Press, 1986), 11–19.
11. McRae, *God Speed the Plough*, 173–4.
12. Cf. Garrett Sullivan, *The Drama of Landscape. Land, Property, and Social Relations on the Early Modern Stage* (Stanford, CA: Stanford University Press, 1998), chapter 1: '"Arden Lay Murdered in that Plot of Ground": Surveying, Land, and *Arden of Faversham*', 31–56, 38.
13. John Norden, *The Surveyors Dialogue* (London: Hugh Astley, 1607), 18. On Norden's career see Frank Kitchen, 'John Norden (c. 1547–1625): Estate Surveyor, Topographer, County Mapmaker and Devotional Writer', *Imago Mundi*, 49 (1997) 43–61.
14. Donald Lupton, *London and the Countrey Carbonadoed* (London: Nicholas Okes, 1632), 106.
15. Norden, *Surveyors Dialogue*, 19–20. Following the Greek geographer Strabo, this was considered to be the historical origin of geometry. See John Dee, 'The Mathematicall Praeface to the Elements of Geometrie of Euclid of Megara' [1570], intr. Allen G. Debus (New York: Science History Publications, 1975), sig. A2r–v, who traces the origins of land surveying back to the same historical precedent.
16. McRae, *God Speed the Plough*, 179.
17. F. M. L. Thompson, *Chartered Surveyors. The Growth of a Profession* (London: Routledge & Kegan Paul, 1968), 1–2.
18. Recorde, *Pathway*, preface.
19. Ibid. (my italics).
20. Which seems to have been popular in mathematical and geographical writing. Cuningham's *Cosmographical Glasse*, Recorde's books, Worsop's (see following note) and Norden's manuals, were all written as dialogues, as was the 1612 treatise *An Olde Thrift Newly Revived*, by 'R. C.' (London: Richard Moore) which features a surveyor as a participant in a debate mainly concerned with agricultural improvement.
21. Edward Worsop, *A Discouerie of sundrie errours and faults daily committed by Landemeaters, ignorante of Arithmetike and Geometrie* (London: Gregorie Seton, 1582), sig. I2v.
22. Ibid.
23. Norden, *Surveyors Dialogue*, 2–6.
24. Thompson, *Chartered Surveyors*, 10.
25. H. C. Darby, 'The Agrarian Contribution to Surveying in England', *Geographical Journal*, 82 (1933) 529–35, 530. Aspects of estate management are still privileged over measuring techniques in Valentine Leigh, *The*

moste profitable and commendable science, of Surueying (London: Miles Jennings, 1577).

26. Aaron Rathborne, *The Surveyor* (London: W. Burre, 1616).

27. Crystal Lynn Bartolovich, 'Spatial Stories: *The Surveyor* and the Politics of Transition', in Alvin Vos (ed.), *Place and Displacement in the Renaissance* (Binghampton, NY: Center for Medieval and Early Renaissance Studies, 1995), 255–83, 260.

28. Historians tend to read the allegory differently. Sarah Bendall, in *Maps, Land, and Society: A History, with a Carto-Bibliography of Cambridgeshire Estate Maps, c. 1600–1836* (Cambridge: Cambridge University Press, 1992), identifies the two figures on the ground as representations of those supposedly ignorant 'fake surveyors' (132) Rathborne never loses an opportunity to deride.

29. Bartolovich, 'Spatial Stories', 273.

30. Cf. Sullivan's discussion of surveying (*The Drama of Landscape*, 39–46) for a similar approach and some suggestive conclusions.

31. Henri Lefebvre, *The Production of Space*, trans. Donald Nicolson-Smith (Oxford: Blackwell, 1991 [French original 1974]), 33, 38–9.

32. Ibid., 39.

33. Dee, 'The Mathematicall Praeface', sig. A2v.

34. Leonard Digges, *A Boke Named Tectonicon* (London: Thomas Marshe, 1566 [first published 1556]).

35. Leonard Digges, *A Geometrical Practise, named Pantometria* (London: Henrie Bynneman, 1571).

36. Cf. E. G. R. Taylor, *Tudor Geography, 1485–1583* (New York: Octagon Books, 1968 [first published 1930]), 91.

37. Darby, 'Agrarian Contribution', 534.

38. Digges, *Tectonicon*, 'L. D. vnto the Reader', n.p. (my italics).

39. Ibid., sig. B1v.

40. John Roche, 'The Cross-Staff as a Surveying Instrument in England 1500–1640', in Sarah Tyacke (ed.), *English Map-Making, 1500–1650. Historical Essays* (London: British Library, 1983), 107–11, 109.

41. Triangulation was first described by Gemma Frisius in an appendix to the 1533 edition of Peter Apian's *Cosmographia*.

42. Radolph Agas, *A Preparative to Platting of Landes and Tenements for Surueigh* (London: Thomas Scarlet, 1596), 9–10.

43. See A. W. Richeson, *English Land Measuring to 1800: Instruments and Practices* (Cambridge, MA: Society for the History of Technology, 1966). For less teleology see the readable account by J. A. Bennett, *The Divided Circle. A History of Instruments for Astronomy, Navigation and Surveying* (Oxford: Clarendon Press, 1987). See also E. G. R. Taylor, 'The Plane-Table in the Sixteenth Century', *Scottish Geographical Magazine*, 45 (1929) 205–11, and the exhibition catalogue by J. A. Bennett and Stephen Johnston, *The Geometry of War, 1500–1750* (Oxford: Museum of the History of Science, 1996).

44. Bennett, *Divided Circle*, 38. The only instrument Fitzherbert mentions is a simple 'Dyall' (*The boke of surueying and improumentes*, sig. I1r), a pocket compass. The first manual to equate surveying with land measuring, Rychard Benese's *The maner of measuring of all maner of lande* (London: James Nicolson, 1537), suggests little more in the way of instruments than the use of a 'pole made of woode' (sig. A2v).

45. Rathborne, *The Surveyor*, 121 (my italics).
46. Another striking example in this context is Arthur Hopton who described his elaborate 'topographical glass' in 1611 as a kind of multipurpose instrument that could serve, according to need, as theodolite, plane table or circumferentor. Arthur Hopton, *Speculum Topographicum: or the Topographicall Glasse* (London: Simon Waterson, 1611).
47. Rathborne, *The Surveyor*, preface (my italics).
48. Worsop, *Discouerie of sundrie errours*, sig. E3r (my italics).
49. The title-page reproduced here as Figure 2.2 is that of the 1592 edition (London: Thomas Orwin).
50. Cyprian Lucar, *A Treatise Named Lucarsolace* (London: Iohn Harrison, 1590).
51. See Skelton and Harvey (eds), *Local Maps and Plans*, passim.
52. P. D. A. Harvey, 'English Estate Maps: Their Early History and Their Use as Historical Evidence', in David Buisseret (ed.), *Rural Images. Estate Maps in the Old and New Worlds* (Chicago: University of Chicago Press, 1996), 27–61, 27. See also Maurice Beresford, *History on the Ground. Six Studies in Maps and Landscape* (London: Lutterworth Press, 1957).
53. Harvey, 'English Estate Maps', 27.
54. P. D. A. Harvey, *Maps in Tudor England* (Chicago: Chicago University Press, 1993), 84.
55. Harvey, 'English Estate Maps', 57.
56. J. B. Harley, 'Meaning and Ambiguity in Tudor Cartography', in Sarah Tyacke (ed.), *English Map-Making*, 22–45, 37.
57. Ibid.
58. William Leybourn, *The Compleat Surveyor* (London: E. Brewster and G. Sawbridge, 1653), 275.
59. Norden, *Surveyors Dialogue*, A5v, 89, A6r.
60. Ibid., 16.
61. Agas, *Preparative*, 14–15.
62. Norden, *Surveyors Dialogue*, 90.
63. For a reading that places such marginal portraits in the context of changing conceptions of ethnic difference on maps of Africa, see my 'Randfiguren. Othello, Oroonoko und die kartographische Repräsentation Afrikas', in Ina Schabert (ed.), *Imaginationen des Anderen im 16. und 17. Jahrhundert* (Wolfenbüttel: Herzog August Bibliothek, forthcoming).
64. Norden, *Surveyors Dialogue*, 28.
65. Ibid., 25.
66. Ibid., 16 (my italics). The same image is used by Worsop, *Discouerie of sundrie errours*, sig. B3v; Lucar, *Treatise Named Lucarsolace*, 53; and Leybourn, *Compleat Surveyor*, 275.
67. McRae, *God Speed the Plough*, 190.
68. Dee, 'The Mathematicall Praeface', sig. A2r–v.

Chapter 3 Surveying Ireland

1. William Folkingham, *Fevdigraphia. The Synopsis or Epitome of Svrveying Methodized* (London: Richard Moore, 1610), title-page.

2. Patricia Coughlan, '"Cheap and Common Animals": The English Anatomy of Ireland in the Seventeenth Century', in Thomas Healy and Jonathan Sawday (eds), *Literature and the English Civil War* (Cambridge: Cambridge University Press, 1990), 205–23, 207.

3. Edmund Spenser, *A View of the Present State of Ireland*, ed. R. F. Gottfried, in *The Works of Edmund Spenser: A Variorum Edition*, eds Edwin Greenlaw et al. (Baltimore, MD: Johns Hopkins Press, 1949), vol. 10: The Prose Works, 40–231. References to the *View* throughout quote line numbers in brackets.

4. John Derricke, *The Image of Irelande, with a discouerie of Woodkarne* (London: John Daie, 1581), sig. D4v, E1r, E3r.

5. This has been J. H. Andrews' object in *Plantation Acres. An Historical Study of the Irish Land Surveyor and His Maps* (Belfast: Ulster Historical Foundation, 1985).

6. The letter is printed in Tomás Ó Laidhin (ed.), *Sidney State Papers, 1565–70* (Dublin: Irish Manuscripts Commission, 1962), 70.

7. On Lythe see J. H. Andrews, 'The Irish Surveys of Robert Lythe', *Imago Mundi*, 19 (1965) 22–31; and his 'Robert Lythe's Petitions, 1571', *Analecta Hibernica*, 24 (1967) 232–41.

8. John Hooker, 'The Chronicles of Ireland', *Holinshed's Chronicles of England, Scotland and Ireland*, 2 vols, 2nd edn (London: Iohn Harrison et al., 1587), vol. 2, 117.

9. I discuss Speed's atlas at length in Chapter 5.

10. Quoted from J. H. Andrews' entry on Lythe in the *Missing Persons* volume of the *DNB* (1993), 422.

11. For an analysis of a later (and possibly better known) mapping project in Ireland, the Ordnance Survey, see Mary Hamer, 'Putting Ireland on the Map', *Textual Practice*, 3, no. 2 (1989) 184–201.

12. Anne Fogarty, 'The Colonization of Language: Narrative Strategies in *A View of the Present State of Ireland* and *The Faerie Queene*, Book VI', in Patricia Coughlan (ed.), *Spenser and Ireland. An Interdisciplinary Perspective* (Cork: Cork University Press, 1989), 76–109, 88. Fogarty draws on an influential essay by Svetlana Alpers, reprinted as 'The Mapping Impulse in Dutch Art', in David Woodward (ed.), *Art and Cartography. Six Historical Essays* (Chicago/London: University of Chicago Press, 1987), 51–96 (esp. 59, 67–9).

13. Cf. Bruce Avery, 'Mapping the Irish Other: Spenser's *A View of the Present State of Ireland*', *ELH*, 57 (1990) 263–79, 270.

14. Such readings frequently draw on a number of seminal essays by J. B. Harley. Cf. especially 'Meaning and Ambiguity in Tudor Cartography', Sarah Tyacke (ed.), *English Map-Making, 1500–1650. Historical Essays* (London: The British Library, 1983), 22–45; 'Maps, Knowledge and Power', in Denis Cosgrove and Stephen Daniels (eds), *The Iconography of Landscape* (Cambridge: Cambridge University Press, 1988), 277–312; 'Silences and Secrecies: the Hidden Agenda of Cartography in Early Modern Europe', *Imago Mundi*, 40 (1988) 57–76; 'Deconstructing the Map', *Cartographica*, 26, no. 2 (1989) 1–20.

15. Julia Reinhard Lupton defines the word 'plot' in this context as 'a phrase encompassing English strategies for Irish reform, the cartographic projects of surveying and mapping which furthered them, and, more generally, a geographical, antiquarian approach to Irish history'. Cf. 'Mapping Mutabil-

ity: or, Spenser's Irish Plot', in Brendan Bradshaw et al. (eds), *Representing Ireland. Literature and the Origins of Conflict, 1534–1660* (Cambridge: Cambridge University Press, 1993), 93–113, 93.

16. That this was an ideological position difficult to achieve in practice is confirmed by extant examples of Irish maps, some of which I examine in Chapter 6 below.

17. For some comments on the conceptual ambiguities in the mapping of the colonial landscape of Ireland cf. David J. Baker, 'Off the Map: Charting Uncertainty in Renaissance Ireland', in Bradshaw et al. (eds), *Representing Ireland*, 76–92.

18. Seamus Heaney, 'Ocean's Love to Ireland', *North* (London: Faber, 1975), 47.

19. Ó Laidhin (ed.), *Sidney State Papers*, 72.

20. Lythe presumably arrived in Ireland on 19 September. Cf. the letter from Thomas Jenison to Cecil, dated Carrickfergus, 25 September 1567, which notes that 'Robert Lythe arrived here six days since' by which time he had already 'drawn two plats of this town.' Letter printed in Robert Dunlop, 'Sixteenth-Century Maps of Ireland', *English Historical Review*, 20 (1905) 309–37, 331.

21. Letter dated 11 June 1567. Printed in Ó Laidhin (ed.), *Sidney State Papers*, 69.

22. On the use of maps in government see two articles by Peter Barber, 'England I: Pageantry, Defense, and Government: Maps at Court to 1550' and 'England II: Monarchs, Ministers, and Maps, 1550–1625', in David Buisseret (ed.), *Monarchs, Ministers and Maps. The Emergence of Cartography as a Tool of Government in Early Modern Europe* (Chicago/London: Chicago University Press, 1992), 26–98.

23. According to Rudolf Gottfried these four garrisons are to '[form] a quadrilateral around Tyrone, they are intended to block all the channels through which outside support might reach the Earl.' Cf. 'Irish Geography in Spenser's *View*', *ELH*, 6 (1939) 114–37, 117. I will return to this point in Chapter 6 which looks more systematically at Irish maps. Gottfried identified the map in question as Baptista Boazio's *Irelande* which has since been dated to 1599 and thus cannot have been known to Spenser. Mercedes Maroto Camino thinks the map in question is 'a detailed military chart' (170) and argues that 'we can only assume that the map of "Ireland" Eudoxus holds, stands for the land and the power Spenser wished to "view" but did not yet possess' (189). See her '"Methinks I See an Evil Lurk Unespied": Visualizing Conquest in Spenser's *A View of the Present State of Ireland*', *Spenser Studies*, 12 (1998) 169–94. I am grateful to Mercedes Camino for providing me with a copy of her essay at proof stage.

24. It is also the region where at least one contemporary surveyor, Richard Bartlett, met his death at the hands of the local population. Cf. George Hayes-McCoy, *Ulster and Other Irish Maps, c. 1600* (Dublin: Stationery's Office, 1964), xii.

25. Lupton, 'Mapping Mutability', 95. Lupton offers what I think is an outstanding analysis of Spenser's cartographics in both the *View* and *The Faerie Queene*.

26. Avery, 'Mapping the Irish Other', 265.

27. I am adapting a phrase from Frank Lestringant who characterizes cartography, with reference to the work of Guillaume Le Testu, as both 'an experience of the world and an experiment on the world' (*Mapping the Renaissance World. The Geographical Imagination in the Age of Discovery* (Cambridge: Polity Press, 1994 [French original 1991]), 104).'

28. Andrew Hadfield, 'Briton and Scythian: Tudor Representations of Irish Origins', *Irish Historical Studies*, 28 (1993) 390–408, 405.

29. Lupton, 'Mapping Mutability', 98.

30. David Norbrook, *Poetry and Politics in the English Renaissance* (London: Routledge & Kegan Paul, 1984), 143.

31. Sir John Davies, *A Discoverie of the trve cavses why Ireland was neuer entirely subdued, nor brought vnder Obedience of the Crowne of England, vntill the Beginning of his Maiesties happie Raigne* (London: Iohn Iaggard, 1612). Further references to the *Discovery* quote page number in brackets.

32. Anne Fogarty has recently analysed Davies' textual strategies as deliberate fabrications in the service of English colonial power: '[I]t may be argued that the trope of inconstancy should be seen as the primary rhetorical device of [the *Discovery*]. Davies' narrative is a network of arguments that constantly shift and undermine themselves and yet insist on sustaining the falsity of their initial premise.' '"This Inconstant Sea-Nimph": History and the Limitations of Knowledge in John Davies' Writings about Ireland', in Timothy P. Foley et al. (eds), *Gender and Colonialism* (Galway: Galway University Press, 1995), 23–34, 31.

33. Hans S. Pawlisch, *Sir John Davies and the Conquest of Ireland. A Study in Legal Imperialism* (Cambridge: Cambridge University Press, 1985), 6.

34. John Speed, *The Theatre of the Empire of Great Britaine* (London: Iohn Sudbury & George Humble, 1611), commendatory verses (Davies' italics).

35. On the policy of assimilation which does not exclude a discourse of Irish difference see Michael Neill, 'Broken English and Broken Irish: Nation, Language, and the Optic of Power in Shakespeare's Histories', *Shakespeare Quarterly*, 45 (1994) 1–32, 4–10.

36. On the multiple ways in which Spenser's Irish exile is inscribed in *The Faerie Queene*, see Andrew Hadfield, *Spenser's Irish Experience: Wilde Fruit and Salvage Soyl* (Oxford: Clarendon Press, 1997).

37. Edmund Spenser, *The Faerie Queene*, ed. A. C. Hamilton (London: Longman, 1977). Subsequent references to *The Faerie Queene* are to this edition.

38. See specifically 1.iv.36, 2.xi.5 and 6.xi.47.

39. Coughlan, 'Cheap and Common Animals', 207.

40. *Paradise Lost*, Book 1, 200–8.

41. See Andrew Hadfield, 'Spenser, Ireland, and Sixteenth-Century Political Theory', *Modern Language Review*, 89, no. 1 (1994) 1–18. Hadfield reads the Faunus episode as an 'allegory of the English presence in Ireland' (16).

42. The legal showdown between Jove and Mutabilitie is staged in a landscape modelled on a mountain range some 15 miles east of Spenser's Irish residence at Kilcolman Castle. See Patricia Coughlan, 'The Local Context of Mutabilitie's Plea', *Irish University Review*, 26 (1996) 320–41, 326–7.

43. Cf. Lupton, 'Mapping Mutability', 109–10.

44. See Chapter 8 below.

45. John Norden, The *Surveyors Dialogue* (London: Hugh Astley, 1607), 6.

Part II Cartographies

1. Jorge Luis Borges, 'Of the Exactitude of Science', *A Universal History of Infamy* (Harmondsworth: Penguin, 1981), 131. Louis Marin comments on this tale in *Utopics: The Semiological Play of Textual Spaces*, trans. Robert A. Vollrath (Atlantic Highlands, NJ: Humanities Press, 1984), 233–7; and Jean Baudrillard uses it as a point of departure in his 1981 essay 'Simulacra and Simulation', trans. Paul Foss et al., *Selected Writings*, ed. Mark Poster (Cambridge: Polity Press, 1988), 166–84.
2. *The Complete Works of Lewis Carroll*, ed. Alexander Woollcott, 10th edn (London: Nonesuch Press, 1966), 556–7. The map 'on a scale of a *mile to the mile*', which resembles that of Borges' imperial cartographers closely, is used to comic effect in *Sylvie and Bruno Concluded* [1893].
3. John Norden, *The Surveyors Dialogue* (London: Hugh Astley, 1607), 15.
4. Conrad Heresbach, *Foure Bookes of Husbandry*, trans. Barnaby Googe (London: Richard Watkins, 1577), 3r.
5. Cf. Andrew McRae, *God Speed the Plough. The Representation of Agrarian England, 1500–1660* (Cambridge: Cambridge University Press, 1996), chapters 5 and 6.
6. David Fletcher defines map consciousness as 'the propensity on the part of an individual or group to represent some of their topographical knowledge in the form of maps': *The Emergence of Estate Maps. Christ Church, Oxford, 1600 to 1840* (Oxford: Clarendon Press, 1995), 1.
7. Richard Helgerson, *Forms of Nationhood. The Elizabethan Writing of England* (Chicago: Chicago University Press, 1992), chapter 3: 'The Land Speaks', 105–47, 105.
8. John Speed, *The Theatre of the Empire of Great Britaine* (London: Iohn Sudbury & Georg Humble, 1611), preface.
9. Thomas Blundeville, *A Briefe Description of Vniversal Mappes and Cardes* (London: Thomas Cadman, 1589), sig. C4r.
10. Ciaran Brady and Raymond Gillespie, 'Introduction', in Brady and Gillespie (eds), *Natives and Newcomers: The Making of Irish Colonial Society, 1534–1641* (Dublin: Irish Academic Press, 1986), 11–21, 17.
11. William Shakespeare, *Richard II*, II.i.40. I am quoting from Stephen Greenblatt et al. (eds), *The Norton Shakespeare* (New York/London: Norton, 1997).
12. Ibid., II.i.50. In assuming here that the word 'plot' has cartographic significance I do not want to argue that Gaunt's speech as a whole (too well known to need quoting here) is the verbal evocation of a map. Critical opinion is divided on this point. Most recently, Garrett Sullivan has argued that Gaunt's speech is cartographically inspired ('Reading Shakespeare's Maps', *The Drama of Landscape. Land, Property, and Social Relations on the Early Modern Stage* (Stanford, CA: Stanford University Press, 1998), 92–123), John Gillies that it is not ('The Scene of Cartography in *King Lear*', in Andrew Gordon and Bernhard Klein (eds), *Literature, Mapping, and the Politics of Space in Early Modern Britain* [forthcoming]). I incline to Gillies' view who argues that 'the speech works by a combination of deixis ('This') and a procession of symbolic images (throne, isle, earth, seat, etc.) of the nation' and that apart from 'plot', no other word in Gaunt's string of images has any 'necessary implication in sixteenth century cartographic discourse'.

13. The elision of Wales is a more widespread contemporary phenomenon. Humfrey Lluyd may have been a proponent of some form of Welsh nation-alism, but John Speed is more representative of the English perspective when he called his map of England and Wales simply 'The Kingdome of England'. And his atlas of the British Isles – that is, of Great Britain and Ireland – encompassed geographically what Speed referred to in his title as 'Great Britain'. If the distinction between the geopolitical markers 'British' and 'English' (or even 'Britain' and 'England') was differently defined by con-temporaries, I found it difficult to always do justice to it in my own writing. I use the former adjective (which for some contemporaries could include Ireland) wherever the reference is clearly to the island of Great Britain, the latter – when the semantic context is geographical rather than cultural – as referring to a political space that may or may not include Wales. The fre-quent blurring of boundaries in contemporary usage is part of my argument, and the 'nation' I refer to throughout is as imprecisely defined as extant maps suggest it appeared to contemporaries.

14. William Cuningham, *The Cosmographical Glasse* (London: John Day, 1559), 119 (my italics).

15. The most extensive carto-bibliography of early modern printed maps of the British Isles is Rodney W. Shirley, *Early Printed Maps of the British Isles, 1477–1650*, completely revised and updated edition (East Grinstead, West Sussex: Antique Atlas Publications, 1991).

16. Edmund Spenser, *The Faerie Queene*, ed. A. C. Hamilton (London/New York: Longman, 1977), V.xi.39.

Chapter 4   The Whole World at One View

1. Thomas Blundeville, *A Briefe Description, of Universal Mappes and Cardes* (London: Thomas Cadman, 1589), sig. A2v.

2. John Dee, 'The Mathematicall Praeface to the Elements of Geometrie of Euclid of Megara' [1570], intr. Allen G. Debus (New York: Science History Publications, 1975), sig. A4r.

3. For the history of English cartography in the early modern period see R. V. Tooley, *Maps and Mapmakers* (London: B. T. Batsford, 1952); Edward Lynam, 'English Maps and Map-Makers of the Sixteenth Century', *The Mapmaker's Art. Essays on the History of Maps* (London: Batchworth Press, 1953), 50–76; Sarah Tyacke and John Huddy, *Christopher Saxton and Tudor Map-Making* (London: British Library, 1980); Sarah Tyacke (ed.), *English Map-Making 1500–1650: Historical Essays* (London: British Library, 1983); P. D. A. Harvey, *Maps in Tudor England* (Chicago: Chicago University Press, 1993); Peter Barber, 'Les îles Britanniques', in Marcel Watelet (ed.), *Gerardi Mercatoris. Atlas Europae* (Anvers: Bibliothèque des Amis du Fonds Mercator, 1994), 43–77. For an account of the use of maps in sixteenth-century government see two articles by Peter Barber, 'England I: Pageantry, Defense, and Government: Maps at Court to 1550' and 'England II: Monarchs, Ministers, and Maps, 1550–1625', in David Buisseret (ed.), *Monarchs, Ministers and Maps. The Emergence of Cartography as a Tool of Government in Early Modern Europe* (Chicago/London: Chicago University Press, 1992), 26–98. There

are numerous general books on the history of cartography of which the most ambitious project is undoubtedly the multi-volume *History of Cartography*, edited by David Woodward and the late Brian Harley for Chicago University Press. The volume on the early modern period has not yet appeared in print.

4. Benedict Anderson, *Imagined Communities. Reflections on the Origin and Spread of Nationalism* (London: Verso, 1983), *passim*. To read Tudor and early Stuart maps of England/Britain as powerful signs of an emerging nationhood is a point that has been most influentially argued by Richard Helgerson, *Forms of Nationhood. The Elizabethan Writing of England* (Chicago: Chicago University Press, 1992).

5. Brief and useful introductions to the history and purpose of early modern maps as well as the various genres are provided by Helen Wallis (ed.), *Historian's Guide to Early British Maps*, Royal Historical Society Guides and Handbooks No. 18 (London: Royal Historical Society, 1994).

6. Barber, 'England I', 33; Harvey, *Maps*, chapter 1 (title).

7. Thomas Elyot, *The Boke Named the Gouernour* (London: Thomas Berthelet, 1531), fol. 37r.

8. Ibid., fol. 37v. Garrett Sullivan has commented on the juxtaposition in this passage of the private space of the study with the public space of the world: 'the map . . . is consumed in a restrictive social space that it helps to define through the act of being consumed' (*The Drama of Landscape. Land, Property, and Social Relations on the Early Modern Stage* (Stanford: Stanford University Press, 1998), 98). For some further reflections on how 'the space of the map is underwritten by the comfort of the body', exploring the links between map reading and domestic interiors, see Gillies, 'The Scene of Cartography'.

9. John Norden, *The Surveyors Dialogue* (London: Hugh Astley, 1607), 16.

10. Robert Burton, *The Anatomy of Melancholy* (Oxford: Henry Cripps, 1621), 351.

11. For a recent account of 'mapping' as a complex system of knowledge transmission see Denis Cosgrove, 'Introduction: Mapping Meaning', in Cosgrove (ed.), *Mappings* (London: Reaktion, 1999), 1–23.

12. J. B. Harley, 'Meaning and Ambiguity in Tudor Cartography', in Tyacke (ed.), *English Map-Making*, 22–45, 22. Harley's final essays were all concerned with the social, political and ethical implications of modern cartography. Cf. 'Maps, Knowledge and Power', in Denis Cosgrove and Stephen Daniels (eds), *The Iconography of Landscape* (Cambridge: Cambridge University Press, 1988), 277–312; 'Silences and Secrecies: the Hidden Agenda of Cartography in Early Modern Europe', *Imago Mundi*, 40 (1988) 57–76; 'Deconstructing the Map', *Cartographica*, 26, no. 2 (1989) 1–20. See also the replies to the last piece collected in *Cartographica*, 26, nos 3 and 4 (1989) 89–127; and Harley's response, 'Cartography, Ethics and Social Theory', *Cartographica*, 27, no. 2 (1990) 1–23.

13. John Norden, *Speculi Britanniae Pars: A Topographical and Historical Description of Cornwall* [1610] (London: Christopher Bateman, 1728), sig. A1r.

14. Discussed more fully in Part III below.

15. William Vaughan, *The Spirit of Detraction* (London: George Norton, 1611), 129.

16. Dee, 'Mathematicall Praeface', sig. A4r. On the wide range of Dee's geographical activities and interests, which included his role as adviser to the Muscovy Company, see E. G. R. Taylor, *Tudor Geography, 1485–1583* (New York: Octagon, 1968 [first published 1930]); and William Sherman, *John Dee: The Politics of Reading and Writing in the English Renaissance* (Amherst: University of Massachusetts Press, 1995).

17. George Owen, *The Description of Pembrokeshire* [1603], ed. Henry Owen (London: J. Clark, 1892), 2–3.

18. It should be noted, though, that county chorographies of this nature may well have been a favourite place to complain about unjust treatment on the part of the central government. Owen's comment shows that such issues were increasingly thought of in cartographic terms but I doubt that it furnishes sufficient evidence to argue that contemporaries in general were slow to grasp the concept of cartographic scale (for which it is widely quoted: see, for example, Victor Morgan, 'The Cartographic Image of "The Country" in Early Modern England', *Transactions of the Royal Historical Society*, 5, no. 29 (1979) 129–54, 138; Howard Marchitello, 'Political Maps: The Production of Cartography and Chorography in Early Modern England', in Margaret J. M. Ezell and Katherine O'Brien O'Keeffe (eds), *Cultural Artifacts and the Production of Meaning. The Page, the Image, the Body* (Ann Arbor: University of Michigan Press, 1994), 13–40, 20–1). John Norden anticipated Owen's complaint when he argued that it is impossible to apply the same scale to every map in an English county atlas since 'some Shires of the greatest magnitude, will require two sheetes of paper Royall, when some other will not containe 1/4 of a sheete.' See *Nordens Preparatiue to his Speculum Britanniae* (London: n.p., 1596), 13.

19. Simonds D'Ewes, *A Compleat Journal of the Votes, Speeches and Debates, both of the House of Lords and House of Commons throughout the whole Reign of Queen Elizabeth, of Glorious Memory* (London: Jonathan Robinson et al., 1693), 169.

20. Bruce Avery, 'Mapping the Irish Other: Spenser's *A View of the Present State of Ireland*', *ELH*, 57 (1990) 263–79, 270.

21. Arthur Hopton, *Speculum Topographicum: or the Topographicall Glasse* (London: Simon Waterson, 1611), sig. A2v (my italics).

22. Abraham Ortelius, *The Theatre of the Whole World* (London: Iohn Norton and Iohn Bill, 1606), preface.

23. Denis Wood, *The Power of Maps* (London: Routledge, 1993), 4–5.

24. Blundeville, *Briefe Description*, sig. C4r.

25. Ortelius, *Theatre*, preface (my italics).

26. Samuel Daniel, *A Defence of Ryme* (London: Edward Blount, 1603), sig. G.4.r.

27. Edward Worsop, *A Discouerie of sundrie errours and faults daily committed by Landemeaters, ignorante of Arithmetike and Geometrie* (London: Gregorie Seton, 1582), sig. B4r.

28. Aaron Rathborne, *The Surveyor* (London: W. Burre, 1616), 207 (my italics).

29. Ibid., n.p.

30. Valentine Leigh, *The Moste Profitable and Commendable Science, of Surueying* (London: Andrew Maunsell, 1577), sig. I1r (chapter title).

31. Norden, *Surveyors Dialogue*, sig. A6r.

32. Radolph Agas, *A Preparative to Platting of Landes and Tenements for Surueigh* (London: Thomas Scarlet, 1596), 12 (my italics).
33. Worsop, *Discouerie of sundrie errours*, sig. D1v.
34. Norden, *Surveyors Dialogue*, 16.
35. In using the term landscape here I endorse Garrett Sullivan's view who argues for an understanding of landscape not merely as the traditional theme of what he terms 'the landscape arts' but as 'one among many possible conceptions of social relations mediated by land' (*The Drama of Landscape*, 4).
36. My comments here only skim the surface of this scene; they do not do justice to it as a whole nor are they so intended. For substantial readings of 'Lear's Map' see Frederic T. Flahiff, 'Lear's Map', *Cahiers Elisabethains* no. 30 (1986) 17–33; Terence Hawkes, 'Lear's Maps', *Meaning by Shakespeare* (London: Routledge, 1992); Sullivan, 'Reading Shakespeare's Maps', *The Drama of Landscape*, 92–123; Gillies, 'Scene of Cartography'. Quotations from *Lear* are to the Folio text in Stephen Greenblatt et al. (eds), *The Norton Shakespeare* (New York and London: Norton, 1997).
37. Hawkes, 'Lear's Maps', 121.
38. Gillies, 'Scene of Cartography'.
39. Gillies confirms this: 'It is just possible that these geographic divisions were visibly displayed to the audience in the form of a stage wall map' (ibid.).
40. Cf. lines 42 and 106.
41. Francis Barker, *The Culture of Violence. Essays on Tragedy and History* (Manchester: Manchester University Press, 1993), 3.
42. Hawkes, 'Lear's Maps', 123.

Chapter 5 Mapping the Nation

1. J. B. Harley and Kees Zandvliet, 'Art, Science, and Power in Sixteenth-Century Dutch Cartography', *Cartographica*, 29, no. 2 (1992) 10–19, 14.
2. Peter Barber, 'A Tudor Mystery: Laurence Nowell's Map of England and Ireland', *The Map Collector*, 22 (1983) 16–21, 19. The 'Gough' Map, named after its eighteenth-century owner Richard Gough and now at the Bodleian Library, Oxford, is a fourteenth-century map of Britain that considerably influenced later maps of the British Isles, including the 1569 Mercator version.
3. The map is part of a notebook which also includes a map of Sicily. It is described by Peter Barber in 'The Minister Put His Mind on the Map', *The British Museum Society Bulletin*, 43 (1983) 18–19.
4. Letter printed in Sir Henry Ellis (ed.), *Original Letters of Eminent Literary Men of the Sixteenth, Seventeenth and Eighteenth Centuries* (London: Camden Society, 1843), 21–3.
5. Steven G. Ellis, *Tudor Frontiers and Noble Power. The Making of the British State* (Oxford: Clarendon Press, 1995), ix.
6. Robin Flower, 'Laurence Nowell and the Discovery of England in Tudor Times', *Proceedings of the British Academy*, 21 (1935) 47–73, 60. The identification is accepted by Thomas Hahn, 'The Identity of the Antiquary Laurence Nowell', *English Language Notes*, 20 (1982/3) 10–18, 14; Barber, 'Tudor

Mystery', 18; Carl T. Berkhout, 'The Pedigree of Laurence Nowell the Antiquary', *English Language Notes*, 23 (1985/6) 15–26, 16. As far as I can see, the
identification is merely conjectural. The initials 'L.N.' underneath the Greek
inscription in the bottom left-hand corner indicate Nowell's authorship but
do not reveal the identity of the reclining figure. Barber suggests that Nowell
and Cecil, then in their mid-30s and mid-40s respectively, correspond in
terms of age to the portraits on the map but the age of neither figure can
be estimated with any precision (Barber, 'Tudor Mystery', 20). And neither
does Nowell's letter to Cecil, considered below, confirm this assumption
(even though it establishes beyond doubt that Nowell and Cecil discussed
matters pertaining to the mapping of Britain).

7. 'Spero profecto me . . . regionem nostram non modo simul universam sed et
    partes ejus omnes, et singulas provincias ita depicturum', Ellis (ed.), *Origi-
    nal Letters*, 21–3 (my translation).

8. Nowell's cartographic skills are evident from a series of regional outline
    maps of Great Britain and Ireland preserved in the British Library (MS
    Cotton Domitian xviii). The presence on these maps of a cartographic grid
    supports the hypothesis that such a grid was once also inscribed on the
    General Description, evidence of which survives only on a 1927 photograph
    of the map. Nowell's expertise in mapping is praised by his contemporary
    Thomas Randolph, of Christ Church, Oxford, in a letter to Cecil dated 25
    April 1562: 'When we [Nowell and Randolph] were both schollers in paris
    and he at that tyme partaker of that small thynge that I had, he travayled
    by the helpe of some Scottyshe men that I dyd acquaynte hym wth, to sette
    forthe the marches betwene Englande and Scotlande: yf nowe yt myghte be
    thoughte worthe his travaile better opportunitie servethe than ever it dyd,
    and I know that he can do yt well.' (See *Calendar of State Papers, Foreign
    Series. Elizabeth, 1561–2*, ed. Joseph Stevenson (London: Longmans et al.,
    1866), item 1051, 631.) Quoted from Pamela N. Black, 'Some New Light on
    the Career of Laurence Nowell the Antiquary', *The Antiquaries Journal*, 62
    (1982) 116–23, 118.

9. The small, vellum-covered notebook containing Nowell's map bears a hand-
    written inscription by Lord Shelburne on its front cover, stating that 'Ld
    Burleigh carried this map always about him.'

10. Cf. Flower, 'Laurence Nowell', 63; A. L. Rowse, *The England of Elizabeth. The
    Structure of Society* (London: Macmillan, 1962 [first published 1950]), 35;
    Barber, 'Tudor Mystery', 20.

11. In the appendix to *Saxton's Survey of England and Wales*, intr. and ed. R. A.
    Skelton, *Imago Mundi*, supplement No. 6 (Amsterdam: Nico Israel, 1974),
    Skelton translates the passage in question (see note 7) as follows: 'I hope
    indeed . . . to depict our country (regionem) both as a whole and in all its
    parts, and also the several counties (provincias)' (16). Since the issue hinges
    on the translation of 'provinicias' it is important to note that the word
    appears here for the second time and that earlier in his letter Nowell uses
    it clearly with reference to England: '. . . eos qui huc usque Angliae
    describendae provinciam susciperunt, tibi usque quaque non satisfecisse'
    (Ellis (ed.), *Original Letters*, 22), which Skelton translates thus: '. . . those who
    have hitherto undertaken to describe the country [that is, Anglia] have not
    in all respects satisfied you [Cecil]' (*Saxton's Survey*, 15). This earlier passage

may imply that Nowell, when using 'provincias' a second time, is not referring to counties but to the 'provinces' of England, Scotland, Ireland and
perhaps Wales. Of course, given the semantic indeterminacy of contemporary Latin usage, it is not conclusive evidence that Nowell did not have
counties in mind when he used 'provincias' again. It should be a warning,
though, not to bend the passage too swiftly to our purposes.

12. Available in a modern reprint as *Christopher Saxton's 16th Century Maps: the
Counties of England and Wales*, intr. William Ravenhill (Shrewsbury:
Chatsworth Library, 1992). For an early biography of Saxton see Sir George
Fordham, *Christopher Saxton of Dunningley. His Life and Work* (Leeds: John
Whitehead & Son, 1928).

13. The analogies between estate and national surveys are instructive. The
extant pass in which Saxton was appointed to survey and map Wales, dated
10 July 1576, makes specific mention that local JPs should appoint 'honest
men' to accompany the cartographer, as well as 'a horseman that can speke
both Welshe and Englishe to safe conduct him to the next market towne'.
Acts of the Privy Council of England, new series, vol. 9: 1575–1577, ed. John
Roche Dasent (London: HMSO, 1894), 159. Modern surveying techniques,
applied successfully to English estates, clearly lent themselves to the
mapping of a space as large as the nation but such a project was accompanied by similar fears of a recalcitrant rural population.

14. Cf. for example Sarah Tyacke and John Huddy, *Christopher Saxton and Tudor
Map-Making* (London: British Library, 1980), 30: 'With the publication of
Ortelius' atlas *Theatrum Orbis Terrarum* in 1570, which was the first attempt
to print a collection of maps of the world to a consistent format, the possibility of doing the same for a kingdom became obvious.'

15. Cf. Victor Morgan, 'Lasting Image of the Elizabethan Era', *Geographical
Magazine*, 52 (1980) 401–8.

16. They are partly reproduced in ibid., 404–5.

17. Denis Wood, The *Power of Maps*, (London: Routledge, 1993), 112, 124–5.
Wood identifies the 'tectonic code' as one of the five essential codes of cartographic intrasignification. I use Wood's semiotic model for the analysis of
maps as a theoretical framework throughout much of the present chapter.

18. Benedict Anderson, *Imagined Communities. Reflections on the Origin and Spread
of Nationalism* (London: Verso, 1983), *passim*.

19. Richard Helgerson, *Forms of Nationhood. The Elizabethan Writing of England*
(Chicago: Chicago University Press, 1992), chapter 3: 'The Land Speaks',
105–47.

20. Wood, *Power of Maps*, 112.

21. I am assuming, for the sake of the argument in this paragraph, that all
editions of Saxton's atlas corresponded in layout to the British Library
copy of the atlas, press-mark C.7.c.1 (from which Plate 6 is reproduced).
I am well aware that this was not the case but believe the argument is
valid nevertheless. For a list of extant copies of Saxton see the appendix
in Ifor M. Evans and Heather Lawrence, *Christopher Saxton, Elizabethan
Map-Maker* (Wakefield, West Yorkshire: Wakefield Historical Publications,
1979).

22. George Owen, The *Description of Pembrokeshire* [1603], ed. Henry Owen
(London: J. Clark, 1892), 2.

23. William Boelhower, 'Inventing America: A Model of Cartographic Semiosis', *Word and Image*, 4, no. 2 (1988) 475–97, 479.

24. Victor Morgan, 'The Cartographic Image of "The Country" in Early Modern England', *Transactions of the Royal Historical Society*, 5, no. 29 (1979) 129–54, 153.

25. Which came in 20 separate sheets and was entitled 'Britannia Insularem in Oceano Maxima'. It has been republished in facsimile with an extensive introduction by R. A. Skelton (*Saxton's Survey*).

26. John Gregory, *Gregorii Posthuma* (London: Laurence Sadler, 1649), 319.

27. John Norden, *An Intended Guyde, for English Travailers* (London: Edward Allde, 1625), preface, n.p. (my italics).

28. Wood, *Power of Maps*, 72.

29. The wall map includes in the top right-hand corner a 'fretwork cartouche . . . [which] contains a summary of English history from Caesar to the Tudors', and in the left-hand margin two cartouches wich 'contain, respectively, an account of the legal systems and courts of England . . . and a geographical description of the country'. Skelton, intr., *Saxton's Survey*, 11.

30. Peter Heylyn, *Microcosmvs, or a Little Description of the Great World* (Oxford: Iohn Lichfield and Iames Short, 1621), 11.

31. William Camden, *Britannia*, trans. Philemon Holland (London: George Bishop & John Norton, 1610), 'The Avthor to the Reader' (my italics). Cf. also Chapter 7 below.

32. Denis Cosgrove, 'Introduction: Mapping Meaning', Cosgrove (ed.), *Mappings* (London: Reaktion, 1999), 1–23, 2.

33. Saxton's maps were in the making when the first edition of the Chronicles was compiled. In the introduction Holinshed included a reference to Seckford, Saxton's patron: 'vnderstanding of the great charges and notable enterprice of that worthie Gentleman maister Thomas Sackeforde in procuring the Chartes of the seuerall prouinces of this Realme to be sette forth, wee are in hope that in tyme he will deliniate this whole lande so perfectly, as shal be comparable or beyonde any deliniation heretofore made of any other region'. Raphael Holinshed, *The Firste volume of the Chronicles of England, Scotlande, and Irelande* (London: Iohn Harrison, 1577), 'Epistle Dedicatorie'.

34. A fine reproduction of the 1616 Latin edition of the *Theatre*, mysteriously subtitled 'a Tudor Atlas', is available as *The Counties of Britain* (London: British Library, 1988).

35. John Speed, *The Theatre of the Empire of Great Britaine* (London: Iohn Sudbury & George Humble, 1611), dedication.

36. Helgerson, *Forms of Nationhood*, 105–47.

37. Louis Marin, *Utopics: The Semiological Play of Textual Spaces*, trans. Robert A. Vollrath (Atlantic Highlands, NJ: Humanities Press, 1984), 210.

38. Ibid.

39. Cyprian Lucar, *A Treatise Named Lucarsolace* (London: Iohn Harrison, 1590), 51–2.

40. See Günter Schilder, 'The Development of Decorative Borders on Dutch Wall Maps Before 1619', *Monumenta Cartographica Neerlandica III* (Alphen aan den Rijn: Canaletto, 1990), 115–46.

41. Cf. Valerie Traub, 'Mapping the Global Body', in Peter Erickson and Clark Hulse (eds), *Early Modern Visual Culture* (Philadelphia: University of Pennsylvania Press, expected 2000).

42. Jurij Lotman, *The Structure of the Artistic Text* [1971], trans. Ronald Vroon, Michigan Slavic Contributions 7 (Ann Arbor: University of Michigan Press, 1977), 237–9. I owe this reference to Manfred Pfister.

43. For an approach to early modern English cartography and chorography that focuses specifically on 'the narrativizing of the world' (32) see Marchitello, 'Political Maps'.

44. This point has been most influentially theorized by Homi K. Bhabha. See his editorial introduction and his essay 'DissemiNation' in *Nation and Narration* (London: Routledge, 1990).

45. Cf. Marin, *Utopics*, 205.

46. Harley and Zandvliet, 'Art, Science, and Power', 14.

47. Michel Foucault, 'Questions on Geography', *Power/Knowledge. Selected Interviews and Other Writings 1972–1977*, ed. Colin Gordon (New York: Pantheon Books, 1980), 63–77, 73–4.

Chapter 6 The Image of Ireland

1. Robert Beale, 'A Treatise of the Office of a Councellor and Principall Secretarie to her Majestie' [1592], annexed to the first volume of Conyers Read, *Mr Secretary Walsingham and the Policy of Queen Elizabeth*, 3 vols (Oxford: Clarendon Press, 1925), vol. 1, 423–43, 428–9.

2. So called because the entire land was to be put *down* on paper for the first time.

3. J. H. Andrews, 'Geography and Government in Elizabethan Ireland', in N. Stephen and R. E. Glasscock (eds), *Irish Geographical Studies in Honour of E. Estyn Evans* (Belfast: Queen's University, 1970), 178–91, 180. Despite my occasional criticism of his views, my understanding of Irish maps is greatly indebted to the erudite scholarship of J. H. Andrews who has almost single-handedly written the history of Irish cartography of the early modern period and beyond.

4. I am following George Hayes-McCoy's hypothesis who identifies as Richard Bartlett the 'one Barkeley' mentioned in the following passage of a letter by Sir John Davies to Salisbury, 28 August 1609: 'though the country be now quiet and the heads of greatness gone, yet [our] geographers do not forget what entertainment the Irish of Tyrconnell gave to a map-maker about the end of the late great rebellion; for one Barkeley being appointed by the late Earl of Devonshire to draw a true and perfect map of the north parts of Ulster (the old maps being false and defective), when he came into Tyrconnell, the inhabitants took off his head, because they would not have their country discovered.' *Calender of State Papers, Ireland, James I., 1608–10*, eds C. W. Russell and John P. Prendergast (London: Longman et al., 1874), 280. See George Hayes-McCoy, *Ulster and Other Irish Maps, c. 1600* (Dublin: Stationery's Office, 1964), xii.

5. Andrews, 'Geography and Government', 180.

6. Beale, 'Treatise', 429.

7. Mercedes Camino notes that even a 'cursory review of [existing carto-graphic and chorographic descriptions of Ireland] soon reveals that there were as yet [end of the sixteenth century] no detailed maps of the whole of Ireland; that those maps were deployed either for strategic or military pur-poses or for the allocation of landed states or for both reasons; [and] that the efforts of English surveyors in Ireland were hindered by local opposi-tion'. ' "Methinks I See an Evil Lurk Unespied": Visualizing Conquest in Spenser's *A View of the Present State of Ireland', Spenser Studies*, 12 (1998) 169–94, 178.

8. Lord Burghley's library contained at least two collections of Irish maps: 'Among Burghley's "Bookes remaining at the Court" after his death, . . . two cartographic items are listed under the head "Written Bookes": a "Booke of Mappes of Ireland in large" and another "Book of Mappes of Ireland in Colours." ' R. A. Skelton and J. Summerson, *A Description of Maps and Archi-tectural Drawings in the Collection Made by William Cecil, First Baron Burghley, Now at Hatfield House* (Oxford: Roxburghe Club, 1971), 15. This volume is particularly valuable for its large-scale reproductions of selected manuscript maps in Burghley's possession.

9. Edward Lynam, 'English Maps and Map-Makers of the Sixteenth Century', *The Mapmaker's Art. Essays on the History of Maps* (London: Batchworth Press, 1953), 50–76, 76.

10. General information on early modern maps of Ireland is provided by Andrews, 'Geography and Government', his exhibition catalogue *Ireland in Maps* (Dublin: Trinity College, 1961), the opening chapters of his book on the Irish land surveyor, *Plantation Acres. An Historical Study of the Irish Land Surveyor and His Maps* (Belfast: Ulster Historical Foundation, 1985) and most recently his *Shapes of Ireland. Maps and Their Makers, 1564–1839* (Dublin: Geography Publications, 1997). Andrew Bonar Law's *Printed Maps of Ireland* (Morristown, NJ: Eagle Press, 1993), contains a very useful typology of early modern maps of Ireland; see also his more recent *The Printed Maps of Ireland 1612–1850* (Dublin: Neptune Gallery, 1997). Robert Dunlop's 'Sixteenth Century Maps of Ireland', *English Historical Review*, 20 (1905) 309–37, though dated and partly incomplete, is still of value.

11. The phrase is Svetlana Alpers'. See her 'The Mapping Impulse in Dutch Art', in David Woodward (ed.), *Art and Cartography. Six Historical Essays* (Chicago/London: University of Chicago Press, 1987), 51–96.

12. Hesiod, *Works and Days, Theogony, The Shield of Herakles*, trans. Richard Lat-timore [1959], (Ann Arbor: University of Michigan Press, 1991), 29.

13. The Greek inscriptions are identified by Robin Flower, 'Laurence Nowell and the Discovery of England in Tudor Times', *Proceedings of the British Academy*, 21 (1935) 47–73, 62. For the most recent account of Nowell's life see the entry on Nowell by Carl T. Berkhout in Helen Damico et al. (eds), *Medieval Scholarship. Biographical Studies on the Formation of a Discipline. Vol. 2: Lit-erature and Philology* (New York/London: Garland, 1998), 3–17. For more spe-cific historical information on the map see Peter Barber, 'A Tudor Mystery: Laurence Nowell's Map of England and Ireland', *The Map Collector*, 22 (1983) 16–21, and his 'The Minister Put His Mind on the Map', *The British Museum Society Bulletin*, 43 (1983) 18–19.

14. I owe this point to Eckhard Lobsien.

15. Michael Neill, 'Broken English and Broken Irish: Nation, Language, and the Optic of Power in Shakespeare's Histories', *Shakespeare Quarterly*, 45 (1994) 1–32, 3.
16. John Gillies, *Shakespeare and the Geography of Difference* (Cambridge: Cambridge University Press, 1994), 62.
17. It has been suggested that the figures' respective postures and the temporal allusions in the Greek inscriptions are meant to indicate that the time it took Nowell to complete the map after making the original proposal occasioned the displeasure of Cecil who expected an earlier delivery (Barber, 'Tudor Mystery', 20). As will have become clear from my reading, if studied in view of the cultural dimension of the image's geographical content this explanation of the temporal allusions hardly does justice to the map's visual complexity. Besides, the period of one and a half years between Nowell's original proposal (June 1563) and the map's completion (late 1564/early 1565), during which a lot of information had to be compiled from a variety of sources, seems hardly long enough to justify either Cecil's impatience, Nowell's worries or the necessity to explain delay by taking refuge in the authority of Hesiod.
18. J. H. Andrews, 'Baptista Boazio's Map of Ireland', *Long Room*, 1 (1970) 29. In his recent book-length study of Irish maps Andrews continues this line of argument and prefers to discuss not Boazio's map but his sources, the Irish surveys of Robert Lythe. Cf. *Shapes of Ireland*, chapter 3: '"Baptiste's Isle": Baptista Boazio, 1599', 57–88.
19. But one fairly common in cosmographical works as, for instance, the two ominous 'Thevet's islands' in French cosmographer André Thevet's unfinished *Grand Insulaire et Pilotage* (1586/7) indicate.
20. Mary Hamer, 'Putting Ireland on the Map', *Textual Practice*, 3 (1989) 184–201, 184. Hamer's essay deals with a later mapping project in Ireland, the Ordnance Survey.
21. Shakespeare is quoted throughout from Stephen Greenblatt et al. (eds), *The Norton Shakespeare* (New York and London: Norton, 1997).
22. Andrew Hadfield suggests the influence of John Derricke's *Image of Ireland* [1581]. Cf. 'Shakespeare, John Derricke, and Ireland: *The Comedy of Errors*, III, ii, 105–6', *Notes and Queries*, 242, no. 1 (1997) 53–4.
23. Edmund Spenser, *A View of the Present State of Ireland*, ed. R. F. Gottfried, in *The Works of Edmund Spenser: A Variorum Edition*, eds Edwin Greenlaw et al. (Baltimore, MD: Johns Hopkins Press, 1949), vol. 10: The Prose Works, 40–231, ll. 3061–2.
24. For a historical contextualization of the Irish references in this play see Christopher Highley, *Shakespeare, Spenser, and the Crisis in Ireland* (Cambridge: Cambridge University Press, 1997), 40–66. Highley argues that 'Shakespeare weaves into the play a provocative and deeply conflicted analysis of the threats and stereotypes associated with Ireland' (42).
25. Andrew Murphy, 'Shakespeare's Irish Wars', *Literature and History*, 5, third series, no. 1 (1996) 38–59, 46.
26. John Hooker, 'The Chronicles of Ireland', *Holinshed's Chronicles of England, Scotland and Ireland*, 2 vols, 2nd edn (London: Iohn Harrison et al., 1587), vol. 2, 77.

27. See David J. Baker, '"Wildehirissheman": Colonialist Representation in *Henry V*', *English Literary Renaissance*, 22 (1992) 37–61; and his *Between Nations: Shakespeare, Spenser, Marvell, and the Question of Britain* (Stanford, CA: Stanford University Press, 1997). Cf. also Highley, *Spenser, Shakespeare, and Ireland*, 145–7.

28. After *Henry V*, references to Ireland virtually disappear from Shakespeare's plays (brief mentions in *Macbeth* and *Henry VIII* excepted). Andrew Hadfield thinks that this silence can be explained with 'the restrictions which were placed on the stage representation of English history after June 1599.' *Literature, Travel, and Colonial Writing in the English Renaissance, 1545–1625* (Oxford: Clarendon Press, 1998), 225. Hadfield goes on to argue that some of Shakespeare's later plays, notably *Othello*, can be read 'as displaced allegories of Irish events' (226).

29. Spenser, *View*, l. 14.

30. Cf. Murphy, 'Shakespeare's Irish Wars', 43.

31. Extant copies of Speed's Irish map frequently show an original colouring which foregrounds either the 32-county division, a relatively recent administrative accomplishment, or the four historical provinces.

32. The mantle is extensively denounced in Spenser, *View*, ll. 1555–651. See also Ann Rosalind Jones and Peter Stallybrass, 'Dismantling Irena: The Sexualizing of Ireland in Early Modern England', in Andrew Parker et al. (eds), *Nationalisms and Sexualities* (New York/London: Routledge, 1992), 157–71.

33. See Hayden White, 'The Forms of Wildness: Archaeology of an Idea', *Tropics of Discourse* (Baltimore, MD/London: Johns Hopkins University Press, 1978), 150–82.

34. Murphy, 'Shakespeare's Irish Wars', 38.

35. I am referring here to the set of campaign maps only rediscovered in the late 1950s and now held by the National Library of Ireland (Ms 2656). High-quality reproductions were published in 1964 with an excellent historical commentary by George Hayes-McCoy to which my understanding of these maps is heavily indebted. The name of Bartlett is also connected with a set of more conventional maps of Ulster, now held by the PRO at Kew. Cf. Hayes-McCoy, *Ulster and Other Irish Maps*, introduction. Bartlett has not attracted much critical commentary. For a recent article (which came to my attention only after my own chapter had been submitted to the press) see Mercedes Camino, '(Un)folding the Map of Early Modern Ireland: Spenser, Moryson, Bartlett, and Ortelius', *Cartographica*, 34, no. 4 (1997) 1–17.

36. Hayes-McCoy, *Ulster and Other Irish Maps*, 8.

37. Raymond Gillespie, 'Thomas Raven and the Mapping of the Claneboy Estates', *Journal of the Bangor Historical Society*, 1 (1981) 7–9, 7.

38. Cf. T. W. Moody, *The Londonderry Plantation 1609–41* (Belfast: W. Mullan & Son, 1939), 191–204; and D. A. Chart (ed.), *Londonderry and the London Companies 1609–1629, Being a Survey and Other Documents Submitted to King Charles I by Sir Thomas Phillips* (Belfast: HMSO, 1928).

39. Thomas Raven to Sir Thomas Phillips, 24 June 1621; Chart (ed.), *Londonderry*, 51.

40. Moody, *Londonderry Plantation*, 194.

41. Cf. Sarah Bendall, 'Estate Maps', in Helen Wallis (ed.), *Historian's Guide to Early British Maps*, Royal Historical Society Guides and Handbooks no. 18 (London: Royal Historical Society, 1994), 37–9.

42. Raven's maps remained in manuscript until published by HMSO (see Chart (ed.), *Londonderry*). Two original copies of the whole collection survive, held by the Lambeth Library (Carew MS 634) and the Drapers' archives. Tracings taken from the originals at Lambeth are held by the PRO Northern Ireland (from which the illustrations here are reproduced). I wish to thank Gail Pollock and her colleagues at Environment Service, Historic Monuments and Buildings, Belfast, for first directing me to the Raven maps and for their generous assistance during the early stages of my research.

43. In his biography of Edmund Campion – an English Jesuit who wrote a history of Ireland in 1571 – Evelyn Waugh referred to Campion's Irish exile in a stunning euphemism as the 'cosy, colonial world of the Pale; no hate, no bloodshed'. *Edmund Campion*, 3rd edn (London: Hollis & Carter, 1961 [first published 1935]), 33.

44. Cf. T. W. Moody et al. (eds), *A New History of Ireland, Vol. III: Early Modern Ireland, 1534–1691* (Oxford: Clarendon Press, 1976), 174–6.

45. Cf. Julia Reinhard Lupton, 'Mapping Mutability: or, Spenser's Irish Plot', in Brendan Bradshaw et al. (eds), *Representing Ireland. Literature and the Origins of Conflict, 1534–1660* (Cambridge: Cambridge University Press, 1993), 93–115.

46. *A Valediction: Of Weeping*, ll. 10–13.

Part III Narratives

1. Thomas Blundeville, *Blundeville His Exercises* (London: John Windet, 1595), title-page.

2. This wall-map, entitled *Nova et Exacta Terrarum Orbis Geographicae et Hydrographicae*, was printed in Amsterdam and/or Antwerp in 1592. The only surviving example (now in the Colegio del Corpus Cristi Valencia) is reproduced in Rodney Shirley, *The Mapping of the World. Early Printed World Maps, 1472–1700* (London: New Holland Publishers, 1993), 191–202. The map was republished in London in 1595 but no extant copy is known.

3. Blundeville, *Exercises*, 245r.

4. Ibid., 249v, 253r.

5. The poem is part of an entertainment written for the queen, printed in Walter Bourchier Devereux, *Lives and Letters of the Devereux, Earls of Essex, in the Reigns of Elizabeth, James I., and Charles I., 1540–1646*, 2 vols (London: John Murray, 1853), vol. 2, 502.

6. Shakespeare, *Richard II*, II.i.40. All Shakespeare quotations are to Stephen Greenblatt et al. (eds), *The Norton Shakespeare* (New York/London: Norton, 1997).

7. Virgil, *Eclogues*, 1, 66.

8. Lesley B. Cormack, '"Good Fences Make Good Neighbours": Geography as Self-Definition in Early Modern England', *Isis*, 82 (1991) 639–61, 640.

9. Devereux, *Lives and Letters*, vol. 2, 502.

10. John Norden, *Specvlvm Britanniae. The first parte. An historicall, & choro-graphicall discription of Middlesex* (London: n.p., 1593), 4.
11. According to Samuel Edgerton, this is a feature characteristic of all maps 'where a people believe themselves divinely appointed to the center of the universe'. 'From Mental Matrix to *Mappamundi* to Christian Empire: the Heritage of Ptolemaic Cartography in the Renaissance', in David Woodward (ed.), *Art and Cartography. Six Historical Essays* (Chicago: Chicago University Press, 1987), 10–49, 26.
12. See my opening paragraph to Chapter 1 above.
13. John Donne, 'A Sermon Preached to the Honourable Company of the Virginian Plantation, 13 November 1622', in George R. Potter and Evelyn M. Simms (eds), *The Sermons of John Donne*, 10 vols (Berkeley/Los Angeles: University of California Press, 1962), vol. 4, 264–82, 280.
14. Shakespeare, *Richard III*, II.iv.52–3 (my italics).

Chapter 7   Imaginary Journeys: Describing Britain

1. William Cuningham, *The Cosmographical Glasse* (London: John Day, 1559), fol. 6–7 (my italics). Cf. also the illustrations from Peter Apian's *Cosmographia* (Figure 3).
2. Frank Lestringant, *Mapping the Renaissance World. The Geographical Imagination in the Age of Discovery* (Cambridge: Polity Press, 1994 [French original 1991]), 2–3. On chorography, see also Lestringant's 'Chorographie et Paysage à la Renaissance', in Yves Giraud (ed.), *Le Paysage à la Renaissance* (Fribourg: Editions Universitaires, 1988), 9–26; and Lucia Nuti, 'Mapping Places: Chorography and Vision in the Renaissance', in Denis Cosgrove (ed.), *Mappings* (London: Reaktion, 1999), 90–108.
3. Lucia Nuti comments that the difference between geography and chorography was thought of in terms of 'intellectual and mathematical' versus 'pictorial and sensual knowledge'. 'Mapping Places', 91.
4. A. L. Rowse, *The England of Elizabeth. The Structure of Society* (London: Macmillan, 1962 [first published 1950]), 31 (my italics).
5. Ibid., 31–2.
6. Making it an 'icon' in C. S. Peirce's typology of signs.
7. Cyprian Lucar, *A Treatise Named Lucarsolace* (London: Iohn Harrison, 1590), 51–2.
8. William Lambarde, *A Perambulation of Kent* (London: Ralphe Newberie, 1576), 18.
9. John Leland, *The laboryouse Journey & serche . . . for Englandes Antiquitees* (London: John Bale, 1549), sig. D4r–v.
10. See D. R. Woolf, *The Idea of History in Early Stuart England* (Toronto, etc.: University of Toronto Press, 1990), 19.
11. See, for instance, Ranulph Higden's *Polychronicon* (fourteenth century). On medieval descriptions of England see Lesley Johnson, 'The Anglo-Norman Description of England: An Introduction', in Ian Short (ed.), *Anglo-Norman Anniversary Essays*, Occasional Publication Series, no. 2 (London: Anglo-Norman Text Society, 1993), 11–30. Cf. also her 'Etymologies, Genealogies, and Nationalities (Again)', in Simon Forde et al. (eds), *Concepts of National*

Identity in the Middle Ages, Leeds Texts and Monographs, New Series 14 (Leeds: Leeds University, 1995), 125–36.

12. Annabel Patterson, *Reading Holinshed's Chronicles* (Chicago/London: University of Chicago Press, 1994), 61.

13. The full title of Harrison's treatise is 'An Historicall Description of the Islande of *Britayne*, with a briefe rehearsall of the nature and qualities of the people of *Englande*, and of all such commodities as are to be found in the same' (my italics). In other words, in a move characteristic of early modern geographical and political thought, the text progresses from historical Britain to contemporary England. In my own analysis, I will concentrate principally on what Harrison has to say about the latter, except where the context requires otherwise.

14. William Harrison, 'An Historicall Description of the Islande of Britayne', Raphael Holinshed (ed.), *The Firste volume of the Chronicles of England, Scotlande, and Irelande* (London: Iohn Harrison, 1577), 20r.

15. Ibid., 'Epistle Dedicatorie'.

16. Ibid., 36r.

17. Ibid., 35v.

18. Michel de Certeau, *The Practice of Everyday Life* [1974], trans. Steven Rendall (Berkeley et al.: University of California Press, 1988), 119.

19. Harrison, 'Historicall Description', 56r.

20. Andrew McRae, *God Speed the Plough. The Representation of Agrarian England, 1500–1660* (Cambridge: Cambridge University Press, 1996), 234.

21. Louis Marin, *Utopics: The Semiological Play of Textual Spaces*, trans. Robert A. Vollrath (Atlantic Highlands, NJ: Humanities Press, 1984), 202.

22. Stan A. E. Mendyk, *'Speculum Britanniae.' Regional Study, Antiquarianism, and Science in Britain to 1700* (Toronto et al.: University of Toronto Press, 1989), 51.

23. William Camden, *Britannia*, trans. Philemon Holland (London: George Bishop & John Norton, 1610), 182.

24. Ibid., 'The Avthor to the Reader', n.p.

25. Richard Helgerson, *Forms of Nationhood. The Elizabethan Writing of England* (Chicago: University of Chicago Press, 1992), 116.

26. Richard Helgerson, 'Nation or Estate? Ideological Conflict in the Early Modern Mapping of England', *Cartographica*, 30, no. 1 (1993) 68–74, 73.

27. Most county chorographies remained in manuscript but some notable printed works are John Norden's *Description of Middlesex* (1593) and *Description of Hertfordshire* (1598); Richard Carew's *Survey of Cornwall* (1603); and William Burton's *Description of Leicestershire* (1622). John Stow's *Survey of London* (1599), though nominally on a city, follows many of the conventions of a county chorography, down to the representation of streets as rivers. Manuscripts would include, for instance, William Smith's *Visitacion of Lancashire* (1598), Thomas Beckham's *Collections for the County of Suffolk* (1602), Sampson Erdeswicke's *Survey of Staffordshire* (completed 1603), George Owen's *Description of the County of Pembroke* (1603), Henry Petowe's *Description of the Countie of Surrey* (1611), Robert Reyce's *Breviary of Suffolk* (1618), Thomas Gerard's *Particular Description of Somerset* (1633) and John Coker's *Survey of Dorsetshire* (1633). Devon could boast three antiquarian accounts: Tristam Risdon, *Chorographical Description or Survey of Devon*

(1630), Thomas Westcote, *A View of Devonshire* (1630) and Sir William Pole, *Collections towards a description of the county of Devon* (1635).

28. Laurence Nowell, however, apparently travelled as far as Ireland. Carl T. Berkhout notes that in June 1560 he 'went by way of Holyhead to Ireland. His travel notes end with his setting out for Waterford on 14 July.' See Berkhout's entry on Nowell in Helen Damico et al. (eds), *Medieval Scholarship. Biographical Studies on the Formation of a Discipline. Vol. 2: Literature and Philology* (New York/London: Garland, 1998), 3–17, 6.

29. Harrison, 'Epistle Dedicatorie'.

30. Helgerson, 'Nation or Estate?'

31. John Norden, *Nordens Preparatiue to his Speculum Britanniae* (London: n.p., 1596), title-page.

32. Ibid., 18.

33. Ibid., 17.

34. Ibid., 20–1.

35. John Norden *Specvlvm Britanniae. The first parte. An historicall, & chorographicall discription of Middlesex* (London: n.p., 1593), dedicatory address to Cecil (my italics).

36. Norden, *Preparatiue*, 1 (my italics). Norden repeated this designation in an address to James upon the new monarch's accession: '[I] have been employed by authority in the re-description of the shires of England'. Quoted from the introduction to John Norden, *Speculi Britanniae Pars: An Historical and Chorographical Description of the County of Essex* [1594], ed. Sir Henry Ellis (London: Camden Society, 1840), xxxv.

37. Quentin Skinner, 'Moral Ambiguity and the Renaissance Art of Eloquence', *Essays in Criticism*, 44, no. 4 (1994) 267–92.

38. Norden, *Preparatiue*, 27 (my italics).

Chapter 8 The Poetics of National Space

1. John Norden, *An Intended Guyde, for English Travailers* (London: Edward Allde, 1625), n.p. (on the table for Yorkshire). The booklet evidently proved quite popular. It was republished a decade later (after Norden's death) as *A Direction for the English Traviller* (London: Mathew Simons, 1635), in an edition that included thumbnail sketch maps of each county.

2. Norden, *Intended Guyde*, n.p. (preface).

3. In choosing epic as the epithet for both poems I mean to emphasize their interaction with the actual politics of their historical moment (as distinguished from the timeless and unspecific geography of romance). My intention is not to engage in a discussion over the distinction between epic and romance, or even to tackle the question of which category best fits Spenser's poem. For the standard discussion of the difference between epic and romance in Western literature see Erich Auerbach, *Mimesis. Dargestellte Wirklichkeit in der abendländischen Literatur*, 8th edn (Bern and Stuttgart: Francke, 1988 [first published 1946]).

4. See Joan Grundy, *The Spenserian Poets* (London: Edward Arnold, 1969).

5. Michael Drayton, *Poly-Olbion* [1612/22], eds William Hebel et al., *The Works of Michael Drayton*, 4 vols (Oxford: Basil, Blackwell, & Mott, 1961 [corrected

edition, with revised bibliography by Bent Juel-Jensen]), vol. 4, sig. v*. I rely on this edition throughout. Further references quote song and line number in brackets.

6. Ibid.

7. Ibid., 391.

8. Jean Brink, *Michael Drayton Revisited* (Boston: Twayne Publishers, 1990), ix. For an earlier study see Joseph A. Berthelot, *Michael Drayton* (New York: Twayne Publishers, 1967).

9. Andrew McRae, *God Speed the Plough. The Representation of Agrarian England, 1500–1660* (Cambridge: Cambridge University Press, 1996), 255.

10. Barbara C. Ewell, 'Drayton's *Poly-Olbion*: England's Body Immortalized', *Studies in Philology*, 75, no. 3 (1978) 297–315, 298–9.

11. The pun is Paul Gerhard Buchloh's. Cf. his *Michael Drayton. Barde und Historiker – Politiker und Prophet. Ein Beitrag zur Behandlung und Beurteilung der nationalen Frühgeschichte Großbritanniens in der englischen Dichtung der Spätrenaissance* (Neumünster: Karl Wachholtz, 1964), 36.

12. Richard Helgerson, *Forms of Nationhood. The Elizabethan Writing of England* (Chicago and London: Chicago University Press, 1992), 108–24, offers a suggestive discussion of Drayton's frontispiece, situating it in the context of the title-page gracing the works of Saxton, Camden and Speed. Arguably, Helgerson's emphasis on the cover runs the danger of disregarding the text itself: '[T]oo often in criticism of *Poly-Olbion* we are asked to take the frontispiece as a synecdoche for the poem's essence ... rather than to read the poem itself.' Claire McEachern, *The Poetics of English Nationhood* (Cambridge: Cambridge University Press, 1996), chapter 4: 'Putting the "poly" back into *Poly-Olbion*', 138–91, 167.

13. Drayton renews his intention to include Scotland in the 1622 preface. Scholars are divided as to whether Drayton gave up this ambitious plan in recognition of the growing lack of interest in his work or whether ill health and other personal difficulties kept him from pursuing his scheme to the end. For the first view see Richard F. Hardin, *Michael Drayton and the Passing of Elizabethan England* (Lawrence et al.: University Press of Kansas, 1973): 'No one can take seriously [Drayton's] promise of "going on with Scotland"' (66); the second is the position of Jean Brink, *Michael Drayton*, who speculates about the poem's 'unfinished conclusion' (95).

14. Alastair Fowler, 'The Beginnings of English Georgic', in Barbara K. Lewalski (ed.), *Renaissance Genres. Essays on Theory, History, and Interpretation* (Cambridge, MA.: Harvard University Press, 1986), 105–25, 119.

15. John Taylor, *Taylor on Thame Isis* (London: John Haviland, 1632), 'To any Body'.

16. Helgerson, *Forms of Nationhood*, 146.

17. McRae, *God Speed the Plough*, 253, 257.

18. Helgerson, *Forms of Nationhood*, 144.

19. See especially Grundy, *The Spenserian Poets*, and Ewell, 'Drayton's *Poly-Olbion*'. Claire McEachern, in *The Poetics of English Nationhood*, thinks differently and argues that *Poly-Olbion*, 'despite its own encouragement to envisage Britain as one giant and harmonious country house, ... offers nothing if not a picture of disintegration' (167).

20. To my knowledge Wyman H. Herendeen has been alone in devoting any critical energy to a reading of the maps rather than merely noting their existence in passing. See *From Landscape to Literature. The River and the Myth of Geography* (Pittsburgh: Duquesne University Press, 1986), esp. 294.

21. Cf. ibid., 294.

22. William Hole, who also engraved the frontispieces of both the *Britannia* and of *Poly-Olbion*.

23. The reference is to C. S. Peirce's semiotic theory which distinguishes between three types of signs: the symbol – the product of convention, characterized by an arbitrary relation between sign and referent; the icon – a sign linked to its referent in a relationship of structural similarity; and the index – where a relationship of proximity – that is, a metonymy – exists between sign and referent.

24. John Selden, 'From the Author of the Illustrations', *Poly-Olbion*, sig. viii* (his italics).

25. See Anne Lake Prescott, 'Marginal Discourse: Drayton's Muse and Selden's Story', *Studies in Philology*, 88, no. 3 (1991) 307–28.

26. Herendeen, *From Landscape to Literature*, 247. Internal quote is from *The Faerie Queene*, IV.xi.18.

27. Jonathan Goldberg, *Endlesse Worke. Spenser and the Structures of Discourse* (Baltimore, MD and London: Johns Hopkins University Press, 1981), 135.

28. Camden's river marriage poem *de Connubio Tamae et Isis* appears in fragments throughout the *Britannia*. An earlier treatment of the river motif had been Leland's *Cygnea Cantio* in the *Itinerary*.

29. Goldberg, *Endlesse Worke*, 135.

30. Harry J. Berger, 'Two Spenserian Retrospects: the Antique Temple of Venus and the Primitive Marriage of Rivers', *Revisionary Play. Studies in the Spenserian Dynamics* (Berkeley et al.: University of California Press, 1988), 195–214, 210.

31. Ibid., 211.

32. From the Spenser/Harvey correspondence first published in 1580. Reprinted in Edmund Spenser, *Poetical Works*, eds J. C. Smith and E. de Selincourt (Oxford: Oxford University Press, 1991 [first published 1912]), 609–43, 612.

33. Cf. 'The Ruines of Times', *Poetical Works*, 473: '*Cambden* the nourice of antiquitie, / And lanterne vnto late succeeding age, / To see the light of simple veritie, / Buried in ruines, through the great outrage / Of her owne people, led with warlike rage. / *Cambden*, though time all moniments obscure, / Yet thy iust labours euer shall endure' (ll. 169–75).

34. Spenser, *Poetical Works*, 600–2.

35. Michael Ondaatje, *The English Patient* (London: Picador, 1993), 261.

36. Gareth Roberts, *The Faerie Queene*, Open Guides to Literature (Buckingham and Philadelphia: Open University Press, 1992), 77. I should point out that Roberts' book is excellent in both its approach and its discussion of the poem.

37. James Nohrnberg, *The Analogy of* The Faerie Queene (Princeton, NJ: Princeton University Press, 1976), 326.

38. John Erskine Hankins, *Source and Meaning in Spenser's Allegory. A Study of 'The Faerie Queene'* (Oxford: Clarendon Press, 1971), 60. My brief typology largely follows Hankins' argument.

39. See Judith H. Anderson, ' "The Hard Begin": Entering the Initial Cantos', in David Lee Miller and Alexander Dunlop (eds), *Approaches to Teaching Spenser's Faerie Queene* (New York: Modern Language Association, 1994), 41–8. My reading of Redcrosse's adventure in Errour's den in this paragraph follows Anderson's who considers the 'episode in the Wandering Wood . . . [as] essentially an exercise in perception' (45).

40. From Latin 'errare', to err and to wander.

41. The most recent effort to describe the geography of *The Faerie Queene* is Wayne Erickson, *Mapping* The Faerie Queene. *Quest Structure and the World of the Poem* (New York and London: Garland, 1996). Erickson argues that the setting of *The Faerie Queene* – partly Fairyland, partly sixth-century Britain, partly the political geography of sixteenth-century Europe – corresponds to the generic multiplicity of Spenser's poetic structure which mixes both epic and romance aspects.

42. This point is supported by Chris Fitter who argues that 'Spenser in *The Faerie Queene* eschews serene and closed landscapes, immune to the incursions of challenging reality, and sets his knights among "open" landscapes whose want of security is underlined.' *Poetry, Space, Landscape. Toward a New Theory* (Cambridge: Cambridge University Press, 1995), 300.

43. Annexed to the first edition of *The Faerie Queene* and reprinted in most modern editions.

44. A similar point has recently been made by Joanne Woolway Grenfell: 'The unchartedness of Faerie land's literary, moral, and geographical territory emphasizes the testing of the knight and reader'. See her 'Do Real Knights Need Maps? Charting Moral, Geographical, and Representational Uncertainty in Spenser's *Faerie Queene*', in Andrew Gordon and Bernhard Klein (eds), *Literature, Mapping, and the Politics of Space in Early Modern Britain* (Cambridge University Press, forthcoming).

45. Radolph Agas, *A Preparative to Platting of Landes and Tenements for Surueigh* (London: Thomas Scarlet, 1596), 6 (wrongly paginated 10).

46. Aaron Rathborne, *The Surveyor* (London: W. Burre, 1616), 168.

47. On the significance of this passage see Annabel Patterson, 'The Egalitarian Giant: Representations of Justice in History/Literature', *Journal of British Studies*, 31 (1992) 97–132.

48. Ovid, *Metamorphosis, Englished by G. S.* [George Sandys] (London: William Stansby, 1626), 5.

49. See my discussion of Saxton and Speed in Chapter 5 above.

50. Homi Bhabha, 'Dissemination. Time, Narrative and the Margins of the Modern Nation', *The Location of Culture* (London: Routledge, 1994), 145.

51. Ibid.

52. The 1592 Ditchley Portrait of Queen Elizabeth, attributed to Marc Gheeraerts the Younger and now in the National Portrait Gallery, London, is a powerful late instance of this idea. But cf. the alternative reading in Helgerson, *Forms of Nationhood*, 112.

Chapter 9 Groundless Fictions: Writing Irish Space

1. Extracts of most relevant contemporary Irish treatises are collected in the following anthologies: Constantia Maxwell (ed.), *Irish History from Contem-*

porary Sources 1509–1610 (London: Allen & Unwin, 1923); J. P. Myers, Jr (ed.), *Elizabethan Ireland. A Selection of Writings by Elizabethan Writers on Ireland* (Hamden, CT: Archon, 1983); Seamus Deane (ed.), *The Field Day Anthology of Irish Writing*, 3 vols (Derry: Field Day Publications, 1991), vol. 1; J. P. Harrington (ed.), *The English Traveller in Ireland. Accounts of Ireland and the Irish through Five Centuries* (Dublin: Wolfhound, 1991); Andrew Hadfield and John McVeagh (eds), *Strangers to That Land. British Perceptions of Ireland from the Reformation to the Famine* (Gerards Cross, Buckinghamshire: Colin Smythe, 1994).

2. Now held by the Public Record Office, Kew, as item MPF 117. The maps that used to accompany the text now have separate callmarks: MPF 65, 66 and 67.

3. Handwritten dedication.

4. Cf. Karl Bottigheimer, 'Kingdom and Colony: Ireland in the Westward Enterprise, 1536–1660', in K. H. Andrews et al. (eds), *The Westward Enterprise. English Activities in Ireland, the Atlantic, and America 1480–1650* (Liverpool: Liverpool University Press, 1978), 45–64.

5. Paul Brown, ' "This thing of darkness I acknowledge mine": *The Tempest* and the Discourse of Colonialism', in Jonathan Dollimore and Alan Sinfield (eds), *Political Shakespeare. Essays in Cultural Materialism*, 2nd edn (Manchester: Manchester University Press, 1994 [first published 1985]), 48–71, 50.

6. See Hayden White's seminal analysis of the 'wild man myth' in 'The Forms of Wildness: Archaeology of an Idea', *Tropics of Discourse* (Baltimore, MD and London: Johns Hopkins University Press, 1978), 150–82. Cf. also Anthony Pagden, *The Fall of Natural Man. The American Indian and the Origins of Comparative Ethnology* (Cambridge: Cambridge University Press, 1982), chapter 1.

7. Andrew Hadfield, ' "The Naked and the Dead": Elizabethan Perceptions of Ireland', in Jean-Pierre Maquerlot and Michèle Willems (eds), *Travel and Drama in Shakespeare's Time* (Cambridge: Cambridge University Press, 1996), 32–54, 48.

8. Andrew Murphy, 'Reviewing the Paradigm: A New Look at Early-Modern Ireland', *Éire-Ireland*, 31, nos 3 and 4 (1996) 13–40, 30. In *But the Irish Sea Betwixt Us. Ireland, Colonialism, and Renaissance Literature* (Lexington: University of Kentucky Press, 1999), Murphy examines images of sameness and difference in the Anglo-Irish encounter by focusing specifically on 'relations of geographic and cultural proximity [which] rendered the Irish, for the English, as "proximate" Others rather than . . . "absolute" Others' (6).

9. Edmund Campion, for instance, declared that though '[t]he contrye is very fruitefull bothe of corne and grasse', it is not properly cultivated, 'for defaulte of housbandry'. *Two bokes of the histories of Ireland* [1571], ed. A. Vossen (Assen: Van Gorcum, 1963), 15.

10. See François Hartog, *The Mirror of Herodotus. The Representation of the Other in the Writing of History*, trans. Janet Lloyd (Berkeley: University of California Press, 1988).

11. Sir John Davies, *A Discoverie of the trve cavses why Ireland was neuer entirely subdued, nor brought vnder Obedience of the Crowne of England, vntill the Beginning of his Maiesties happie Raigne* (London: Iohn Iaggard, 1612), 4–5.

12. Cf. especially the passage in Spenser's *View* where Irenius suggests the 'sword' as the principal strategy of political reform, explaining that 'all those evills muste firste be Cutt awaie by a stronge hande before anie good Cane be planted, like as the Corrupte braunches and vnholsome boughes are firste to be pruned and the foul mosse clensed and scraped awaye before the tree cane bringe forthe anye good fruite.' Edmund Spenser, *A View of the Present State of Ireland*, ed. R. F. Gottfried, in *The Works of Edmund Spenser: A Variorum Edition*, eds Edwin Greenlaw et al. (Baltimore, MD: Johns Hopkins University Press, 1949), vol. 10: The Prose Works, 40–231, ll. 2956–60. The passage has been suggestively analysed by Eamon Grennan, 'Language and Politics: A Note on Some Metaphors in Spenser's *View of the Present State of Ireland*', *Spenser Studies*, 3 (1982) 99–110.

13. Luke Gernon, 'A Discourse of Ireland' [MS c. 1620], included in Caesar Litton Falkiner, *Illustrations of Irish History and Topography, Mainly of the Seventeenth Century* (London et al.: Longmans & Co., 1904), 348–62, 349.

14. Which remained in print until 1628. Cf. Chapter 1 above.

15. Gernon, 'Discourse of Ireland', 349.

16. Ibid., 349–50.

17. Ibid., 350, 353, 355.

18. Louis Montrose, 'The Work of Gender in the Discourse of Discovery', *Representations*, 33 (1991) 1–41, 5.

19. For an exemplary analysis of this image see ibid., 1–7. Montrose's essay focuses on another well-known text that makes use of this trope, Sir Walter Ralegh's *The Discoverie of the Large, Rich, and Beautifull Empire of Guiana* (1596), which contains the infamous description of South American Guyana as 'a countrey that hath yet her maydenhead'. Richard Hakluyt, *The Principal Navigations, Voyages, Traffiques and Discoveries of the English Nation*, 12 vols (Glasgow: James MacLehose & Sons, 1904), vol. 10, 428.

20. Gernon, 'Discourse of Ireland', 350 (my italics).

21. This critical paradigm is principally associated with the work of Irish historians David Beers Quinn and Nicholas Canny. See especially Quinn's seminal study of Anglo-Irish contacts in the Elizabethan period, *The Elizabethans and the Irish* (Ithaca, NY: Cornell University Press, 1966), which contains a chapter called 'Ireland and America Intertwined'; and Canny's influential analysis of English rule in sixteenth-century Ireland, *The English Conquest of Ireland: A Pattern Established* (Hassocks, Sussex: Harvester, 1976). While the exceptional quality of Quinn's and Canny's scholarship remains beyond doubt, some of their premises appear increasingly problematic, particularly the personal links Canny sought to establish when he spoke of some American colonists' 'years of apprenticeship' in Ireland (*English Conquest*, 159). For some incisive remarks on an 'Early Modern Ireland [that] has been dragged into mid-Atlantic, cut off from Europe and left an English stepping-stone to North America' (50), see Hiram Morgan, 'Mid-Atlantic Blues', *Irish Review*, no. 11 (1991/2) 50–6. Cf. also Hadfield, 'The Naked and the Dead', and Murphy, *But the Irish Sea Betwixt Us*, 1–32.

22. It is important to point out in this context that Spenser's allegory for Ireland in *The Faerie Queene*, Irena, may be read – like most of the female characters in the poem, if Spenser's letter to Ralegh is taken seriously – as an

Elizabeth figure. The implications of such a reading – inevitably a critique of Elizabeth's rule in Ireland – are examined by Andrew Hadfield in *Spenser's Irish Experience: Wilde Fruit and Salvage Soyl* (Oxford: Clarendon Press, 1997). Hadfield describes Irena as 'a cipher, an empty figure, who stands for a blank Ireland which needs to be represented and defined by the New English colonists' (156).

23. See Montrose, 'Work of Gender', 7–14.
24. For a study that takes issue with the view of Dee as an 'Elizabethan magus' – a critical position associated most closely with the names of Frances Yates and Peter French – and which offers instead an explicitly political reading of Dee's life and writings, see William Sherman, *John Dee: The Politics of Reading and Writing in the English Renaissance* (Amherst: University of Massachusetts Press, 1995). Sherman rightly points out that 'as long as Elizabeth reigned – and for some time after – the British Empire remained an entirely textual affair' (152).
25. Peter Hulme, *Colonial Encounters. Europe and the Native Caribbean, 1492–1797*, 2nd edn (London and New York: Routledge, 1992 [first published 1986]), 1, 159–60.
26. The following passage from Strabo is quoted both in Campion, *Two bokes*, 21; and in William Camden, *Britannia*, trans. Philemon Holland (London George Bishop & John Norton, 1610), 140 (separate pagination in section on Ireland).
27. Strabo, *The Geography*, Loeb Classics, 8 vols, trans. H. L. Jones (Cambridge, MA: Harvard University Press, 1932), vol. 2, 259–60. Cf. also this passage in Diodorus Siculus: 'The most savage peoples among them are those who dwell beneath the Bears and on the borders of Scythia, and some of these, we are told, eat human beings, even as the Britains do who dwell on Iris [Ireland], as it is called.' *The Library of History*, Loeb Classics, 12 vols, trans. C. H. Oldfather (Cambridge, MA: Harvard University Press, 1939), vol. 3, 181.
28. John Hooker, 'Giraldus Cambrensis: The Conquest of Ireland', *Holinshed's Chronicles of England, Scotland and Ireland*, 2 vols, 2nd edn (London: Iohn Harrison et al., 1587), vol. 2, 6.
29. William Arens, 'Rethinking Athropophagy', in Francis Barker et al., *Cannibalism and the Colonial World* (Cambridge: Cambridge University Press, 1998), 39–62, 40.
30. John Hooker, 'The Chronicles of Ireland', *Holinshed's Chronicles*, 2nd edn, vol. 2, 182–3 (my italics).
31. Ibid.
32. Robert Payne, 'A Brief Description of Ireland: Made in the Year 1589', in Aquilla Smith (ed.), *Tracts Relating to Ireland*, 2 vols (Dublin: Irish Archaeological Society, 1841), vol. 1, 13.
33. Campion, *Two bokes*, 93.
34. Hooker, 'Chronicle', 68.
35. Quoted in Canny, *English Conquest*, 126.
36. Fynes Moryson, *An Itinerary* [1617–26], 4 vols (Glasgow: MacLehose, 1907–8), vol. 3, 282 (my italics). The analogy between Moryson's 'some old women' and early modern representations of witchcraft is striking and hardly accidental. Witches, like cannibals, assault the integrity of the human

body and need to be socially marginalized. For the narrow discursive connections between European witches and New World cannibals, both gendered female, see Sigrid Brauner, 'Cannibals, Witches, and Shrews in the "Civilizing Process"', *Mitteilungen des Zentrums zur Erforschung der Frühen Neuzeit*, 2 (1994) 29–54.

37. William Farmer, 'Chronicles of Ireland' [1615], in Hadfield and McVeagh (eds), *Strangers to That Land*, 101–2 (my italics).

38. Hulme briefly discusses this topos (*Colonial Encounters*, 81).

39. For an analysis of how and why the word 'canibal' and its cognates entered the European languages see Hulme, *Colonial Encounters*, chapter 1: 'Columbus and the Cannibals', 13–44.

40. Since it cannot be disproved either, Arens is careful not to claim that anthropophagy has never existed, only that no satisfactory evidence for its existence is known. William Arens, *The Man-Eating Myth: Anthropology and Anthropophagy* (New York: Oxford University Press, 1979). This provocative study which advanced the thesis that the emergence of anthropology as a discipline is dependent on the cultural construction of anthropophagy sparked off an intense and polemical discussion, revisited by Peter Hulme in the introduction to Barker et al., *Cannibalism and the Colonial World*, 'The Cannibal Scene', 1–38. In his own contribution to this volume, 'Rethinking Anthropophagy', Arens repeats his conviction that 'the ever-present cannibals on the horizon of the Western world are the result of intellectual conjuring – including the anthropological variety' (40).

41. Gananath Obeyesekere, 'Cannibal Feasts in Nineteenth-Century Fiji: Seaman's Yarns and the Ethnographic Imagination', in Barker et al. (eds), *Cannibalism and the Colonial World*, 63–86, 63.

42. Arens, *The Man-Eating Myth*, 22.

43. Denis Bethell, 'English Monks and Irish Reform in the Eleventh and Twelfth Centuries', *Historical Studies. Papers Read Before the Irish Conference of Historians*, no. 8 (1971) 115–32, 125; see also John Gillingham, 'The English Invasion of Ireland', in Brendan Bradshaw et al. (eds), *Representing Ireland. Literature and the Origins of Conflict, 1534–1660* (Cambridge: Cambridge University Press, 1993), 24–42. The *locus classicus* for the identification of the Irish as barbarians in the medieval period are the writings of Giraldus Cambrensis. Describing the Irish, Giraldus reports that 'their external characteristics of beard and dress, and internal cultivation of the mind, are so barbarous that they cannot be said to have any culture. . . . This people is, then, a barbarous people, literally barbarous.' Giraldus Cambrensis, *The History and Topography of Ireland*, trans. John J. O'Meara (Harmondsworth: Penguin, 1982), 101.

44. Quinn, *The Elizabethans and the Irish*, 26.

45. Moryson, *Itinerary*, vol. 4, 203.

46. Davies, *Discoverie*, 166–7.

47. Barnabe Rich, *A New Description of Ireland* (London: Thomas Adams, 1610), 18.

48. Arens, 'Rethinking Anthropophagy', 54.

49. Hulme, *Colonial Encounters*, 86.

50. Ibid., 3.

51. Davies, *Discoverie*, 110 *et passim*.

52. Thomas Gainsford, *The Glory of England* (London: Thomas Norton, 1618), 144.
53. Ibid., 145–7.
54. Spenser, *View*, ll. 560–7.
55. Davies, *Discoverie*, 1–2.
56. Spenser, *View*, ll. 3254–64.
57. Though one may note the striking linguistic parallels, down to exact verbal repetitions, with the report on the famine in Ulster given by Campion, *Two bokes* (see note 33). If an intertextual relation exists (and I think it does) it should be pointed out that 25 years have elapsed since Campion's description and that both passages refer to entirely different events.
58. An observation reminiscent of the etymology of 'barbarian', a term initially used to designate all those unable to speak Greek.
59. Spenser, *View*, ll. 3266–70.
60. Stephen Greenblatt, *Marvelous Possessions. The Wonder of the New World* (Oxford: Clarendon Press, 1991), 122.

Index